T0155498

# SpringerBriefs in Energy

More information about this series at http://www.springer.com/series/8903

Claudio Madeddu · Massimiliano Errico
Roberto Baratti

# CO$_2$ Capture by Reactive Absorption-Stripping

## Modeling, Analysis and Design

 Springer

Claudio Madeddu
Dipartimento di Ingegneria Meccanica,
Chimica e dei Materiali,
Università di Cagliari
Cagliari, Italy

Roberto Baratti
Dipartimento di Ingegneria Meccanica,
Chimica e dei Materiali
Università di Cagliari
Cagliari, Italy

Massimiliano Errico
Department of Chemical Engineering,
Biotechnology and Environmental
Technology, University of Southern
Denmark
Odense M, Denmark

ISSN 2191-5520          ISSN 2191-5539   (electronic)
SpringerBriefs in Energy
ISBN 978-3-030-04578-4      ISBN 978-3-030-04579-1    (eBook)
https://doi.org/10.1007/978-3-030-04579-1

Library of Congress Control Number: 2018963280

This Springer imprint is published by the registered company Springer Nature Switzerland AG
The registered company address is: Gewerbestrasse 11, 6330 Cham, Switzerland

# Contents

# Chapter 1
# Introduction

This introductory chapter includes a discussion on the concept of pollution and a description of the relationship between the increase of the carbon dioxide emissions and the pollution of the four spheres of the Earth system, i.e., atmosphere, hydrosphere, geosphere and biosphere. A brief history of climate change is presented from a scientific and a political point of view. The carbon capture and storage technologies, considered as the most effective solutions for the immediate mitigation of the anthropological $CO_2$ emissions, are introduced and described. In the end, the motivations and the objectives of the book are reported, together with a short summary of the different chapters.

## 1.1  Pollution and Carbon Dioxide

The term pollution is an extremely powerful catchword often misused in headlines, political programmes, news or to condemn a significant number of different activities [1]. Nevertheless, a unanimous general definition of this word does not exist [2–4]. This aspect is fundamental in the field of International Law, as observed by Springer [2], in order to give a legal value to this term. A noteworthy number of attempts has been done to produce a general description for pollution [5]. For example, in the Stockholm Declaration on the Human Environment in 1972, pollution was defined as [6]: *the discharge of toxic substances or of other substances and the release of heat, in such quantities or concentrations as to exceed the capacity of the environment to render them harmless*. This definition, though it gives a precise description of "what" causes pollution, lacks to identify "who" is responsible of polluting. This question is somewhat answered by the OECD Recommendation of 1974 regarding "Principles concerning Transfrontier

Pollution" [7]: *pollution means the introduction by man, directly or indirectly, of substances or energy into the environment resulting in deleterious effects of such a nature as to endanger human health, harm living resources and ecosystems, and impair or interfere with amenities and other legitimate uses of the environment.* This appears to be a good compromise for a complete and general definition of pollution, and its importance resides in the fact that the responsibility by man is clearly highlighted. In fact, as reported by Russel [1], it is important to distinguish between natural phenomena, which are unavoidable, and the man activities, where man is expected to have control. However, the typical use of the term pollution involves the association with a certain sphere of the Earth system, i.e., geosphere, biosphere, hydrosphere and atmosphere [3, 4].

Among the extended number of topics regarding the environmental pollution, the one which has probably gained most of the attention at different levels in the recent years is the one concerning the carbon dioxide. This compound has a central role in the plants' photosynthesis process. Moreover, thanks to its presence in the atmosphere, the average global temperature guarantees the existence of life on Earth. This is true when the amount of $CO_2$ in the atmosphere is at its natural level of about 0.03 vol%. When this amount increases, a chain of problems involving all the four Earth spheres rises:

1. **Atmosphere**. Climate change due to greenhouse gases emissions, of which $CO_2$ is the main contributor [8], is considered a global issue. In fact, the average temperature of the earth has risen year by year from the industrial revolution in the second half of the 18th century. This is the first effect of the chain;
2. **Hydrosphere**. The increase of the level of $CO_2$ solubilized in the oceans contributes to the so-called ocean acidification. This effect is proven to have an important impact on marine ecosystems and biodiversity [4];
3. **Geosphere**. The increase of the mean average temperature causes the ice melting, leading to the rise of the sea level. Without any intervention, this fact is going to cause the inundation of land in certain parts of the world [9, 10];
4. **Biosphere**. While reading you have inhaled in a single breath about 0.59 grams of air including 0.365 milligrams of carbon dioxide. Nevertheless, in 1959, you would have breathed in only 0.284 milligrams of it. It is easy to imagine the impact of these numbers on the public opinion and the general fear in the potential relationship between environmental quality and health care spending [11].

The global warming has become one of the major issues in the field of climate change. Thanks to the work of scientists, especially after WWII, this problem has gained the attention not only of the scientific community, but it has also reached policy makers and citizens.

## 1.2  Brief History of Climate Change Science

It was 1824 when Jean-Baptiste Joseph Fourier discovered the greenhouse effect [12]. In particular, he found that, due to the distance from the sun, the Earth should have been much colder than it actually was. For this reason, he stated that *some blanket* in the air stopped the heat radiations from the sun to be reflected back outside the Earth atmosphere, making possible to heat the planet. However, Fourier did not have the necessary equipment to perform any experiment, and for that moment his intuition remained as they were. It was only after almost 40 years that John Tyndall suggested that the water vapor present in the air was the reason why the radiations from the sun were trapped inside the atmosphere [13]. At the same time, he considered the effect of $CO_2$ negligible, due to the fact that it was in considerably less amount compared to water vapor. It was Svante Arrhenius in 1896 to discover that the blanket predicted by Fourier was mainly caused by the presence of carbon dioxide in the air [14]. Arrhenius found a correlation between the average Earth temperature and the amount of carbon dioxide in the atmosphere. In particular, he estimated that halving the amount of $CO_2$ in the atmosphere would have led to a drop of 4–5 K  of the mean temperature of the Earth, while doubling the same amount would have a lead to a rise of about 5–6 K. Moreover, he was the first to recognize that the burning of fossil fuels was one of the main sources for the accumulation of $CO_2$ in the atmosphere. In the following decades, different works corroborated the intuitions by Arrhenius. For example, Callendar, in 1938, found a relation between the increase of carbon dioxide and the increase of the mean global temperature [15]. Successively, in 1956, Plass stated that the so-called "Carbon Dioxide Theory" could explain different climate change related phenomena [16]. He stated that the carbon dioxide emitted by industrial activities at the rate of the '50s would cause the increase of 1.1 K  of the mean global temperature each century. The effect of the $CO_2$ on the Earth temperature was becoming year by year more evident, and the only way to assess this effect was to monitor the concentration of $CO_2$ in the atmosphere by accurate measurements. In 1958 Charles D. Keeling started to measure the $CO_2$ concentration in the atmosphere of Antarctica and Mauna Loa Island (Hawaii), founding in the following years that this parameter was having an increasing trend [17]. Since then, concern about the mean global temperature increase started rapidly to rise in the scientific community. In particular, the emissions of $CO_2$ caused by human activities were definitely highlighted. In 1981, Wigley and Jones stated that *"The effects of $CO_2$ may not be detectable until around the turn of the century. By this time, atmospheric $CO_2$ concentration will probably have become sufficiently high (and we will be committed to further increases) that a climatic change significantly larger than any which has occurred in the past century could be unavoidable."* [18]. This growing concern in the scientific community, although it was an important result, was just a first step towards the search for a solution to the problem of global warming. In fact, the next step was to bring the problem at a political level.

## 1.3 The Climate Change in the Political Debate

Only in the second half of the '80s the scientific community reached an almost unanimous agreement on the necessity to find an answer to the worrying temperature rise due to emissions of $CO_2$ from industrial sources [19, 20]. In particular, an effort in terms of policy making was needed. Of course, this request faced numerous obstacles. In fact, the problem of global warming, together with the environment, was going to greatly affect the economy of different countries, especially those that were the main emitters of the greenhouse gas [19–21]. The first important result was reached at the World Conference on the Changing Atmosphere held in Toronto (Canada) in 1988. In this occasion, it was stated the recommendation to reduce the emissions of $CO_2$ by at least 20% compared to the 1988 levels by 2005 [22]. However, doubts still remained on how to reach this objective. One of the main obstacles to the development of the reduction strategy was due to the opposition of governments who were frightened by the economic consequences of this new trend of reduction of the $CO_2$ emissions. In particular, most of them claimed uncertainties in the actual danger evaluation and produced several reports trying to, in some way, underestimate the problem. Also, they denounce the lack of scientific officiality in the previous reports. For this reason, it was raised the need to form an intergovernmental institution, thanks to which science and politics could be somehow mixed in order to produce official reports and guidelines for policy makers with the aim to find a concrete solution. Then, in 1988, the Intergovernmental Panel on Climate Change (IPCC) was founded. An extensive work on the origin of the IPCC can be found in the works by Agrawala [19, 23]. In 1990 the IPCC First Assessment Report was approved [24]. One of the main results of this first report was that, under the rate of $CO_2$ emissions at the start of the '90s, the mean global temperature was likely to rise by about 0.3 K per decade. Up to now, the IPCC has produced a total of five Assessment Reports.

Together with the IPCC, the '90s saw also the birth in 1992 in Rio de Janeiro (Brazil) of the United Nations Framework Conventions on Climate Change (UNFCC), which is an agreement between countries included among the United Nations members with the purpose of facilitating the intergovernmental climate change negotiations. The most important outcomes of this framework were reached with the well-known Kyoto Protocol (Kyoto-Japan) in 1997, signed during the Third Conference of Parties (COP-3), which imposed the reduction of the emissions of greenhouse gases (carbon dioxide, methane, nitrous oxide, hydrofluorocarbons, perfluorocarbons, sulphur hexafluoride) *"by at least 5% below 1990 levels in the commitment period 2008–2012"* for 37 industrialized countries [25]. However, as reported by Bodansky, although clear obligations were imposed, the specific ways to reach the objective were still under negotiations [21]. More recently, at the COP-21 in Paris (France) in 2015, 194 states and European Union signed the so-called Paris agreement, with the main aim of *"holding the increase in the global average temperature to well below* 2 °C *above pre-industrial levels and to pursue efforts to limit the temperature increase to* 1.5 °C *above pre-industrial levels, recognizing that this would significantly reduce the risks and impacts of climate change*" [26]. Then, it is evident the importance reached by the global warming problem also at a political level. At the same time, as often happens in the political debate, the situation remains quite complex and in continue evolution. For example, in 2017, USA has announced the intention to withdraw from the Paris agreement [27]. This is due to the fact that the global warming, together with the danger for the Earth, comes together with important implications both at an economic and national interests' level. Negotiations are typically slow and not always clear, and each word in every agreement is carefully selected in order to be accepted by the signatories [21].

## 1.4   The Role of the Public Opinion

The last element in the global debate is represented by the public opinion. As reported in different papers, the global warming problem is felt by the citizens with different degrees of concern [28, 29]. Furthermore, the climate change topic has gained a central position in the media and the newspapers [30–33]. The public opinion represents a fundamental factor in the choices at a political level and, consequently, an important component in the actuation of the provisions to solve the problem. It is then important for citizens to be informed and to understand the importance of finding a solution to the global warming in order to have an active role in the global debate. This is because global warming is damaging both the environment and the ecosystems, and it is of great interest of all the countries to get on a sustainable path.

## 1.5   Carbon Capture and Storage Technologies

The purpose to find and implement actions to mitigate the global warming has brought to different reactions at different scientific, political and social levels. The strong correlation between the carbon dioxide emissions and the mean global temperature increase is nowadays one central topic in the global debate. $CO_2$ is produced in every process involving combustion reactions. With the advent of the industrialization, the amount of carbon dioxide derived from anthropological sources has risen continuously year by year. Starting from the birth of the IPCC, different strategies have been proposed to mitigate the problem. Among them, increase the use of renewable energies has always been present and considered mandatory. At the same time, it is unrealistic and, nevertheless, not economically sustainable, to suddenly end the use of fossil fuels. For this reason, it was reported in the IPCC First Assessment Report that "*the technologies to capture and sequester $CO_2$ from fossil fuel combustion deserve investigation, considering the expected continuing dependence on fossil fuels as primary energy sources*" [24]. The requested effort to develop this kind of technologies was also cited in the Art. 2 of the Kyoto Protocol, where an explicit call for "*Research on, and promotion, development and increased use of, new and renewable forms of energy, of carbon dioxide sequestration technologies and of advanced and innovative environmentally sound technologies*" was made [25]. For this reason, Carbon Capture and Storage (CCS) technologies were born with the purpose of mitigating the emissions of $CO_2$ from industrial plants involving combustion processes. Though technologies to capture and store the $CO_2$ have been known since at least the start of the '80s [34–37], the first explicit reference to CCS technologies by the IPCC was made in 2005, when the organization published a special report on Carbon Capture and Storage [38]. In this work, Carbon Capture & Storage was defined as "*a process consisting of the separation of $CO_2$ from industrial and energy-related sources, transport to a storage location and long-term isolation from the atmosphere [...]. The $CO_2$ would be compressed and transported for storage in geological formations, in the ocean, in mineral carbonates, or for use in industrial processes*". It is not a coincidence that, according to Scopus, the number of works related to carbon capture and storage technologies has increased considerably at that point. The importance of CCS has grown rapidly after the start of the new millennium thanks to the work of the scientific community and institutions like the Global CCS Institute and IEA. In its last report in 2017, the Global CCS Institute declared that: "*Currently, the world is way off track in meeting the Paris Agreement climate goals, and it cannot get back on track without CCS.*" and "*International climate change bodies (IPCC, IEA) confirm that CCS is the only mitigation technology able to decarbonise large industrial sectors. CCS and renewables are partner technologies working towards the same decarbonised objective.*" [39]. According to the same report, 17 large scale CCS facilities are currently operating in the world, and four new ones are coming in 2018.

CCS technologies can be divided into three main categories:

- **Pre-combustion Capture**. In this case the $CO_2$ is captured before the combustion process by means of reforming and gasification processes, obtaining the so-called *syngas*, which is a gas mixture of essentially $H_2$ and CO. Together with the $CO_2$ capture, the object of this process is to produce a gas stream with a high hydrogen content which can be used in IGCC (Integrated Gasification Combined Cycle) processes. Due to the high costs, this technology is not diffused;

- **Oxy-Combustion**. In this process the combustion is conducted with pure oxygen instead of air. The final product is an exhaust gas highly concentrated in $CO_2$ which can be immediately captured and stored. The main problem related to this technology is the necessity of a continuous oxygen supply, which leads to high costs.

- **Post-combustion Capture**. $CO_2$ is captured from the exhaust gas generated from the combustion of fuel. In general, carbon dioxide is diluted in the exhaust gas, leading to the necessity to use specific processes for its removal. Different separation alternatives are available such as adsorption, physical absorption, cryogenic separation, membrane absorption or algal systems [40]. Among them, chemical absorption is undoubtedly the most appealing method. The main advantage of this technology relies in the fact that it can be easily integrated in existing plants compared to the other two options [41].

In this book, the $CO_2$ post-combustion capture by means of chemical absorption was chosen as target process, since it is considered the most mature and promising technology for industrial development [40].

## 1.6   Book Motivations and Objectives

The $CO_2$ post-combustion capture by chemical absorption-stripping is a process consisting in the reactive absorption by an aqueous chemical solution and subsequent regeneration of the solvent by means of a reactive stripping process. The process has been studied since the '50s and a very exhausting literature exists on different topics, such as the choice of a proper solvent, the study of the kinetics, the synthesis of different process configurations etc. Among the different research branches, process modeling is one of those that has mostly evolved thanks to the availability of numerous commercial software simulators. During the years, different models have been proposed in the literature, and nowadays the scientific community agrees that the so-called rate-based model is the most indicated one for the rigorous process description in this field. This has been demonstrated through a significant number of works regarding the validation of the model by means of experimental data [42–50]. Thanks to the availability of such models, the scientific community has started to focus on the process design, which is a crucial point in the development of the process at an industrial level.

The objective of this book is to show a systematic procedure for the steady-state model-based design of a $CO_2$ post-combustion capture plant by reactive absorption-stripping using monoethanolamine as solvent. The first part is dedicated to the process modeling in one of the most powerful process simulation software, Aspen Plus®. In particular, a step-by-step description is presented for the definition of the system thermodynamics and the reaction sets, and the set-up of the rate-based model coupled with the analysis of the system fluid dynamics. The purpose of the first part is to show the correct way to use the simulator when a reactive absorption-stripping process needs to be modeled and simulated. Then, the validation of the model for both the absorber and the stripper is presented using experimental data sets from pilot plant facilities. It is shown how the correct use of the software leads to important improvements in the process modeling, particularly for the description of the typical temperature bulge in the absorber and the estimation of the reboiler duty in the stripper. The second part of the book regards the analysis of the design of an industrial scale plant, based on the contemporary focus on the internal profiles in the columns and the final plant performance. The two sections, i.e., absorption and stripping, are discussed separately, highlighting the crucial points for their correct design. In particular, for the absorber the conditions to avoid isothermal zones in the packing are determined. For what concerns the stripper, an alternative plant configuration without reflux is presented together with a new procedure for the determination of the packing height. In the final part, the interconnection between the two sections is taken into account and the flowsheet is completed with the addition of the solvent water-wash section and the auxiliary equipment. The economic analysis of the complete plant is also performed.

## 1.7  Book Overview

The content of the book is here briefly presented.

- **Chapter** 2. This chapter regards the model set-up in Aspen Plus®. In the first part, the thermodynamics and the reactions sets are defined using the Aspen Properties® package. Then, the important parameters of the rate-based model are described and their determination is discussed. Furthermore, the analysis of the system fluid dynamics and the evaluation of the number of segments is reported.
- **Chapter** 3. The model presented in Chapter 2 is validated for the absorber by means of different experimental data sets from two pilot plant facilities. In particular, two cases are presented: one with the bulge at the bottom of the column and one with a column-top bulge.
- **Chapter** 4. This chapter deals with the model validation for the stripper. Two different pilot plant facilities are again considered for what concerns the experimental data. The effect of the choice of different sets of degrees of freedom is studied for the profiles description and for the estimation of the reboiler duty.
- **Chapter** 5. The design of the absorption section is taken into account using the model developed in the previous chapters. The influence of the molar L/V ratio, which affects the amount of solvent to be used in the process, is highlighted by means of the analysis of the liquid temperature profiles.
- **Chapter** 6. The stripping section is considered here. After the introduction of an alternative plant configuration without reflux, the most important operating process parameters are described in detail. Then, the effect of the packing height on the reboiler duty and the column diameter is analyzed and a criterion for the definition of the minimum packing height is proposed.

- **Chapter** 7. In this chapter, the effect of rich solvent temperature and the design of the cross heat-exchanger, which interconnects the absorber and the stripper, are reported in the first place. Then, after the addition of the solvent recovery section and the auxiliary equipment to the flowsheet, the economic evaluation of the plant is performed by means of the Aspen Process Economic Analyzer®.
- **Chapter** 8. A summary of the highlights discussed in the book is reported in this chapter.

# References

1. Russel V (1974) Pollution: concept and definition. Biol Cons 6(3):157–161
2. Springer AL (1977) Towards a meaningful concept of pollution in international law. ICLQ 26 (3):531–557
3. Birnie PW, Boyle AE, Redgwell C (2009) International law and environment. Oxford University Press
4. Shi Y (2016) Are greenhouse gas emissions from international shipping a type of marine pollution? Mar Pollut Bull 113(1–2):187–192
5. van Heijnsbergen P (1979) The "Pollution" concept in international law. Envtl Pol'y and L 5:11–13
6. A/CONF.48/14/Rev.1 (1972) Report of the United Nations conference on the human environment. United Nations
7. Recommendation of the council on principles concerning transfrontier pollution. OECD (1974)
8. Victor DG, Zhou D, Ahmed EHM et al (2014) Chapter 1—introductory chapter. In: Climate change 2014: mitigation of climate change. Technical report IPCC working group III contribution to AR5, Cambridge University Press, Cambridge
9. Rehan Dastagir M (2015) Modeling recent climate change induced extreme events in Bangladesh: a review. Weather Clim Extremes 7:49–60
10. Shaltout M, Tonbol K, Omstedt A (2015) Sea-level change and projected future flooding along the Egyptian Mediterranean coast. Oceanologia 57(4):293–307
11. Apergis N, Gupta R, Lau CKM et al (2018) U.S. state-level carbon dioxide emissions: does it affect health care expenditure? Renew Sust Energ Lev 91:521–530
12. Fourier J-BJ (1827) Mémoire Sur Les Températures Du Globe Terrestre Et Des Espaces Planétaires. Mem Acad Sci Inst France 7:569–604
13. Tyndall J (1863) On Radiation through the Earth's Atmosphere. Lond Edinb Dubl Phil Mag 25(167):200–206
14. Arrhenius S (1896) On the influence of carbonic acid in the air upon the temperature of the ground. Lond Edinb Dubl Phil Mag 41(251):237–276
15. Callendar GS (1938) The artificial production of carbon dioxide and its influence on temperature. Q J Roy Meteor Soc 64(275):223–240
16. Plass GN (1956) The carbon dioxide theory of climatic change. Tellus 8(2):140–154
17. Keeling CD (1960) The concentration and isotopic abundances of carbon dioxide in the atmosphere. Tellus 12(2):200–203
18. Wigley TML, Jones PD (1981) Detecting $CO_2$-induced climatic change. Nature 292:205–208
19. Agrawala S (1998) Context and early origins of the intergovernmental panel on climate change. Clim Change 39(4):605–620
20. Zillman JW (2009) A history of climate activites. WMO Bulletin 58(3):141–150

21. Bodansky D (2001) The history of the global climate change regime. In: International relations and global climate change, The MIT Press, Massachussets Institute of Technology, Cambridge
22. Usher P (1989) World conference on the changing atmosphere: implications for global security. Environment 31(1):25–27
23. Agrawala S (1998) Structural and process history of the intergovernmental panel on climate change. Clim Change 39(4):621–642
24. IPCC (1992) Climate change: the IPCC 1990 and 1992 assessments. IPCC first assessment report overview and policymaker summaries and 1992 IPCC supplement. WMO
25. UNFCCC (1998) Kyoto protocol to the United Nations framework convention on climate change. United Nations
26. UNFCCC (2015) Paris agreement. United Nations
27. Zhang H-B, Dai H-C, Lai H-X et al (2017) U.S. withdrawal from the paris agreement: reasons, impacts, and China's response. Adv Clim Change Res 8(4):220–225
28. Lorenzoni I, Pidgeon NF (2006) Public views on climate change: European and USA perspectives. Clim Change 77(1–2):73–95
29. Hamilton LC (2011) Education, politics and opinions about climate change evidence for interaction effects. Clim Change 104(2):231–242
30. Porter E (2018) Fighting climate change?. We're Not Even Landing a Punch, The New York Times
31. Friedman L, Popovich N, Fountain H (2018) Who's most responsible for global warming? The New York Times
32. Plumer B (2018) Greenhouse gas emissions rose last year. Here are the top 5 reasons. The New York Times
33. Nuccitelli D (2018) California, battered by global warming's weather whiplash, is fighting to stop it. The Guardian
34. Albanese S, Steinberg M (1980) Environmental control technology for atmospheric carbon dioxide. Energy 5(7):641–664
35. Riemer P (1996) Greenhouse gas mitigation technologies, an overview of the $CO_2$ capture, storage and future activities of the IEA Greenhouse Gas R&D programme. Energy Convers Manag 37(6–8):665–670
36. Audus H (1997) Greenhouse gas mitigation technology: an overview of the $CO_2$ capture and sequestration and further activities of the IEA Greenhouse Gas R&D programme. Energy 22 (2–3):217–221
37. Gunter WD, Wong S, Cheel DB et al (1998) Large $CO_2$ sinks: their role in the mitigation of greenhouse gases from an international, national (Canadian) and provincial (Alberta) perspective. Appl Energy 61(4):209–227
38. IPCC (2005) Intergovernmental panel on climate change, special report on carbon dioxide capture and storage. Technical report Cambridge University Press, Cambridge, United Kingdom
39. Global CCS Institute (2017) The Global Status of CCS: 2017. Australia
40. Wang M, Lawal A, Stephenson P et al (2011) Post-combustion $CO_2$ capture with chemical absorption: a state-of- the-art-review. Chem Eng Res Des 89(9):1609–1624
41. Krishna Priya GS, Bandyopadhyay S, Tan RR (2014) Power system planning with emission constraints: effects of CCS retrofitting. Process Saf Environ Prot 92(5):447–455
42. Tontiwachwuthikul P, Meisen A, Jim Lim C (1992) $CO_2$ absorption by NaOH, monoethanolamine and 2-Amino-2-Methyl-1-Propanol solutions in a packed column. Chem Eng Sci 47(2):381–390
43. Tobiesen FA, Svendsen HF (2007) Experimental validation of a rigorous absorber model for $CO_2$ postcombustion capture. AIChE J 53(4):846–865

44. Tobiesen FA, Juliussen O, Svendsen HF (2008) Experimental validation of a rigorous desorber model for $CO_2$ post-combustion capture. Chem Eng Sci 63(10):2641–2656
45. Zhang Y, Chen H, Chen C-C et al (2009) Rate-based process modeling study of $CO_2$ capture with aqueous monoethanolamine solution. Ind Eng Chem Res 48(20):9233–9246
46. Lawal A, Wang M, Stephenson P et al (2009) Dynamic modelling of $CO_2$ absorption for post combustion capture in coal-fired power plants. Fuel 88(12):2455–2462
47. Plaza JM, Van Wagener D, Rochelle GT (2009) Modeling $CO_2$ capture with aqueous monoethanolamine. Energy Procedia 1(1):1171–1178
48. Mac Dowell N, Samsatli NJ, Shah N (2013) Dynamic modelling and analysis of an amine-based post-combustion $CO_2$ capture absorption column. Int J Greenhouse Gas Control 12:247–258
49. Errico M, Madeddu C, Pinna D et al (2016) Model calibration for the carbon dioxide-amine absorption system. Appl Energy 183:958–968
50. Madeddu C, Errico M, Baratti R (2017) Rigorous modeling of a $CO_2$-MEA stripping system. Chem Eng Trans 57:451–456

# Chapter 2
# Process Modeling in Aspen Plus®

In this chapter, the implementation in Aspen Plus® of the $CO_2$ post-combustion capture by reactive absorption-stripping process model is presented. Components and thermodynamics in the Properties Environment are considered in the first place. Then, in the Simulation environment, the set-up of the *RadFrac*™ model—Rate-Based mode, considered mandatory for this kind of process, is extensively described. The attention is especially focused on the appropriate definition of the rate-based model parameters needed for the discretization of the liquid film. A section is dedicated to the examination of the system fluid dynamics by means of the evaluation of the Peclet number and the number of segments analysis. In particular, it is highlighted how this procedure is of fundamental importance to obtain the correct solution of the resulting system of algebraic equations.

## 2.1 Process Description

The $CO_2$ post-combustion capture using amine aqueous solutions consists in a reactive absorption-solvent regeneration process. Two main sections can be identified in the plant: the absorption, where carbon dioxide is transferred from the vapor/gas phase to the liquid one, and the stripping, where the solvent is regenerated. In particular, the absorption process is enhanced by the reaction between the $CO_2$ transferred in the liquid phase and the solvent. On the other hand, the reverse reaction happens in the stripper to detach the amine from the $CO_2$, which is then transferred back to the gaseous phase. The two sections are interconnected by a cross heat-exchanger. A simplified flowsheet of the system is reported in Fig. 2.1.

The flue gas rich in $CO_2$ enters the absorber from the bottom and flows countercurrent with respect to the liquid solvent. Following the absorption process, the exhaust gas exits the top of the column and after solvent recovery is sent to the stack. The rich solvent is pumped from the bottom of the absorber to the cross heat-exchanger where it is heated and then sent to the top of the stripper. In this

© The Author(s), under exclusive license to Springer Nature Switzerland AG 2019
C. Madeddu et al., *CO₂ Capture by Reactive Absorption-Stripping*,
SpringerBriefs in Energy, https://doi.org/10.1007/978-3-030-04579-1_2

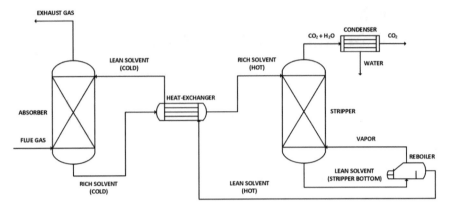

**Fig. 2.1** Simplified flowsheet of a $CO_2$ capture by reactive absorption-stripping plant

second section, the liquid flows countercurrent with the vapor flow generated by the reboiler. From the top of the stripper, a gaseous mixture of carbon dioxide and water is sent to a partial condenser where the $CO_2$ is concentrated in the gas phase and then sent to compression, while the water is recovered in the liquid phase. In this book, different plant configurations have been considered for what concerns the stripper with particular attention to the water flow rate from the condenser, as it is going to be showed in Chap. 6. For this reason, in this chapter no further information is given on the water recovered in the condenser. Exiting from the bottom of the stripper, the regenerated solvent is sent to the heat-exchanger where it supplies its sensible heat to the rich solvent and recycled back to the top of the absorber. Typically, the two columns are packed columns, chosen over the plate ones because the packing is able to provide a higher contact area between the gaseous and the liquid phase and ensures less pressure drop. The presence of the chemical reactions adds complexity to the process, where different phenomena are involved, as schematically reported in Fig. 2.2.

Then, due to the complex nature of the process, a model that is able to take contemporarily into account all of these phenomena is needed in order to obtain a correct mathematical description of the system. In this book, the process was modeled using Aspen Plus® v8.8 and a step-by-step procedure is presented in the following sections.

## 2.2  Building the Model in Aspen Plus®

Aspen Plus® from AspenTech is one of the most used steady-state process modeling and simulation software in process engineering. It is widely used both at an academic and industrial level. It offers the possibility to model a broad spectrum of processes, from the classic distillation columns to different kinds of reactors, and

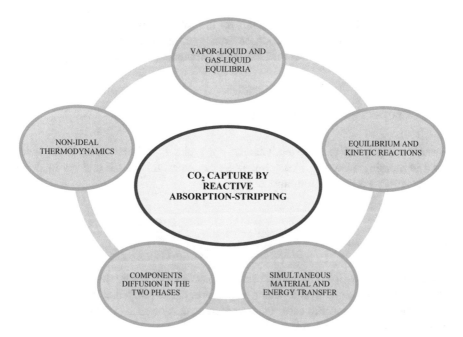

**Fig. 2.2** Phenomena involved in the $CO_2$ capture by reactive absorption-stripping

includes also all the auxiliary equipment typical of a chemical plant, such as valves, pumps, compressors, mixers, and pipelines. The suite offers the possibility to link the modeling part with Aspen Properties®, that gives the possibility to access an extremely large database of physical properties.

As reported in the previous section, the $CO_2$ post-combustion capture by means of amine aqueous solutions consists in a reactive absorption-stripping process. In Aspen Plus® this kind of processes are typically modeled using the so called *RadFrac*™ model, which allows to model absorbers and strippers with chemical reactions. In the following sections the model is described for what concerns:

- Components and thermodynamics;
- Chemical reactions;
- Material and energy balances;
- Interphase transfer;
- Fluid dynamics.

From a mathematical point of view, in Aspen Plus® it is not necessary to write the model equations. In fact, each block contains the material and energy balances used to describe each unit operation. Though this gives the possibility to save time in writing the equations, it rises the necessity to set the model parameters in an appropriate way to ensure the correct process description.

In general, an Aspen Plus® model is built in two distinct environments: Properties (Sect. 2.3) and Simulation (Sect. 2.4). In the first environment, the components are defined and the methods for the computation of the thermodynamic properties are chosen. In the second environment, the streams and the equipment are represented in a flowsheet and the main parameters of the model are set.

## 2.3   Properties Environment

System components and thermodynamics are typically the first inputs to be specified in building an Aspen Plus® model. In particular, this part is entirely made in the Aspen Properties® environment.

### 2.3.1   Components

Two phases are involved in the process: the gaseous and the liquid one. In the first phase, nitrogen, oxygen and water vapor are present together with the $CO_2$. Depending on the specific case, other components such as $H_2S$ can be present as impurities in the flue gas [1].

In the liquid phase, the system $CO_2$-$H_2O$-Amine is characterized by the presence of ions due to ionic dissociation. In this book, monoethanolamine (MEA) was chosen as target solvent, since it is by far the most studied and proven to be the most mature one for this process [2–5]. Once the main components are specified in the *Components—Specifications* panel, Aspen Properties® recognizes the possible presence of ions and asks if the user wants the program to work with one of the following approaches:

- **Apparent components**. In this case the components in the liquid phase are considered undissociated;
- **True components**. In this case for the liquid phase the dissociation of the components in free and dissociated forms is taken into account.

The true components approach is used in this book for the model development to give a more realistic description of the system. Once the option is activated, the software generates all the components involved, including the ions, as reported in Fig. 2.3. For the $CO_2$-MEA-$H_2O$ system, 11 components are generated. Moreover, the software proposes a set of ionic equilibrium reactions and salts dissociation reactions that are used to determine the compositions of all the streams involved in the process. The user has the possibility to choose which reactions are to be used. The salts dissociation reactions are assumed negligible and, subsequently, a set of five chemical equilibrium reactions is considered [6–11]. Figure 2.4 resumes the ionic equilibrium reactions.

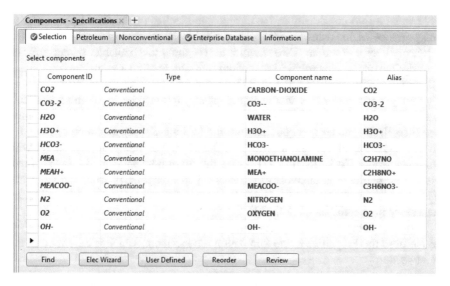

**Fig. 2.3** List of components in the Aspen Properties® —*Components—Specifications* panel

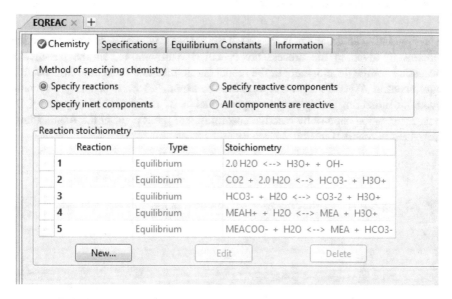

**Fig. 2.4** Equilibrium reactions for the $CO_2$-MEA-$H_2O$ system

## 2.3.2   *Thermodynamics*

After the components are defined, it is possible to specify the model to be used for the evaluation of the thermodynamic properties. In particular, this model must be able to take into consideration the strong non-ideality of the liquid phase due to the presence of ions. In the literature, a significant number of works reports the Electrolyte Non-Random Two Liquid to be the most suitable model for the description of the electrolytic interactions present in the $CO_2$-MEA-$H_2O$ system [2, 6, 8, 10–20]. Then, the Elec-NRTL model was used for the evaluation of the thermodynamic properties in the liquid phase. It must be highlighted that this method is also the one suggested automatically by Aspen Properties®. Furthermore, the Elec-NRTL model was coupled with the Redlich-Kwong Equation of State for the computation of the non-idealities of the vapor/gas phase, in agreement with different works [6, 8, 18, 19, 21]. Once the thermodynamic model is specified, Aspen Properties® automatically retrieves all the parameters from the database. There is also the possibility to change the value of the parameters, which might be necessary in specific cases [14, 22].

After the components and the system thermodynamics are set-up, it is possible to move from the Properties to the Simulation environment, which is described in the next section.

## 2.4 Simulation Environment

The Simulation environment is the part of the software where the equipment and the streams involved in the process are specified. Furthermore, all the parameters necessary for the development of the model are defined in this section. Streams and equipment in Aspen Plus® are reported in the *Main Flowsheet* panel by a specific symbolic notation, giving a clear representation of the process. A typical $CO_2$ post-combustion capture by reactive absorption-stripping plant in the Aspen Plus® flowsheet is reported in Fig. 2.5.

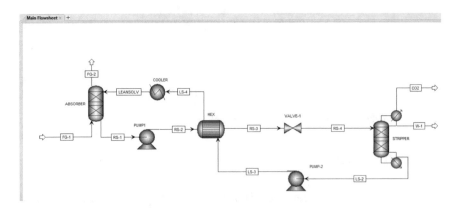

**Fig. 2.5** Aspen Plus® process flowsheet for the $CO_2$ reactive absorption-stripping

As it can be noticed from Fig. 2.5, every piece of equipment is represented by blocks which are linked to each other by arrows representing the streams.

Once the flowsheet is completed, the user must provide a certain number of specifications related to the streams characterization and the equipment features in order to saturate all the degrees of freedom and perform the simulation of the plant. Moreover, if chemical reactions are present in the system, all of them must be reported in the *Reactions* panel.

As already reported in Sect. 2.2, for the modeling of reactive absorption stripping processes, the *RadFrac*™ model is available in Aspen Plus®. The next subsections are dedicated to the set-up of the model.

### 2.4.1 Streams Characterization

The streams are characterized by different parameters. In particular, the user must provide two values among the temperature, the pressure and the vapor fraction. Furthermore, the flow rate must be specified, either in terms of volume, mass or number of moles. For what concerns the component compositions, these can be given in different forms: flow rate, molar or mass fraction, concentration etc. In the case of the $CO_2$ post-combustion capture with amine aqueous solutions, the gaseous phase is straightforward. For the liquid phase, due to the presence of ions, the user has two possibilities:

- Specify the actual composition of each component, both free and ionic forms. In this case the stream composition is defined immediately;
- Specify only the apparent composition of the main components, i.e., $CO_2$, MEA and $H_2O$. In this case the compositions of the ions are determined by Aspen Plus® solving a system of equations that includes the equilibrium reactions defined in the Properties environment (Fig. 2.4), the relations between the apparent and the actual composition [23], and the electro-neutrality equation.

The second possibility is the most common one, since typically the experimental data for the composition are given in the form of apparent compositions. It must be also noted that all the thermodynamic properties of the streams are evaluated using the specifications defined in the Properties environment.

### 2.4.2 Chemical Reactions

Due to the reactive nature of the process, a proper set of reactions must be specified in the *Reactions* panel. As reported in numerous works on the $CO_2$ post-combustion capture with MEA, both kinetic and equilibrium reactions are involved [14–16, 18, 19, 24–30]. In particular, a set including three ionic equilibrium reactions

(Eqs. 2.1–2.3) and two kinetic reversible reactions involving the $CO_2$ (Eqs. 2.4 and 2.5) was considered [18, 19].

$$2H_2O \rightleftharpoons H_3O^+ + OH^- \tag{2.1}$$

$$MEA^+ + H_2O \rightleftharpoons H_3O^+ + MEA \tag{2.2}$$

$$HCO_3^- + H_2O \rightleftharpoons H_3O^+ + CO_3^{2-} \tag{2.3}$$

$$CO_2 + MEA + H_2O \rightleftharpoons MEACOO^- + H_3O^+ \tag{2.4}$$

$$CO_2 + OH^- \rightleftharpoons HCO_3^- \tag{2.5}$$

For what concerns the equilibrium reactions, for the determination of the equilibrium constants as a function of temperature, Aspen Plus® offers two methods:

- **Standard Gibbs free-energy change**. In this case the equilibrium constant has the following rigorous expression (Eq. 2.6):

$$K_{eq} = \exp\left(-\frac{\Delta G^0}{RT^L}\right) \tag{2.6}$$

were the values of the $\Delta G^0$ are retrieved from the Aspen Properties® database.
- **Parameter-based correlation**. In this case the equilibrium constant has the following form (Eq. 2.7):

$$\ln(K_{eq}) = A + \frac{B}{T^L} + C\ln(T^L) + DT^L \tag{2.7}$$

where the coefficients $A$, $B$, $C$, $D$ can be found in different sources [6, 11].

The rigorous approach using the standard Gibbs free-energy change was used in this work, updating the $\Delta G^0$ for the two ions $MEA^+$ and $MEACOO^-$ with the values reported in the AspenTech guideline [31]. For what concerns the expression of the equilibrium constant as function of the components concentration, the activities were chosen as concentration basis.

On the other hand, for kinetic reactions, the default expression in Aspen Plus® is the classic power law, where the kinetic constants are expressed by means of the Arrhenius law (Eq. 2.8):

$$k = k^0\left(-\frac{E_a}{RT^L}\right) \tag{2.8}$$

In particular, the parameters for the evaluation of the kinetic constants for the reaction in Eq. 2.4 were taken from the work of Errico et al. [18], while the parameters for the reaction in Eq. 2.5 were taken from Pinsent et al. [32].

## 2.5   The Aspen Plus® RadFrac™ Model

Once the streams characterization and the reactions are defined, the next step is to describe what happens inside the columns. The first decision to be taken regards the degree of approximation of the model. It was highlighted in the previous sections the complex nature of the process that involves several different phenomena. Then, in order to obtain a correct mathematical description of the system, a model that is able to describe contemporarily the non-ideal thermodynamics, the chemical reactions, the interphase transfer, the component transport in the two phases and the fluid dynamics is needed. The thermodynamics and the reactions were described in Sects. 2.3.2 and 2.4.2, respectively. This section deals with the modeling of the interphase transfer and the components transport in the two phases in the presence of chemical reactions. In the *RadFrac*™ model, two different approaches are present: the equilibrium stages mode and the rate-based mode. In both cases, the column height is discretized into a certain number of parts, which are referred to as *stages* in Aspen Plus®, though in the case of the rate-based mode they should be referred to as *segments* [18, 19].

### 2.5.1   Equilibrium Stages Mode

The equilibrium stages approach assumes that the liquid and the gaseous phases are in intimate contact for a time sufficient for the establishing of the thermodynamic equilibrium between the streams exiting each stage. In the case of the $CO_2$ absorption-stripping, the phase equilibrium assumption is inadequate, due to the contemporary presence of material transfer and chemical reactions. Nevertheless, the equilibrium stages approach can still be used introducing the Murphree efficiency [33], that takes into account the deviation from the phase equilibrium, as it was done in several works [8, 34–36]. As reported in different works, the typical values of the Murphree efficiency for the reactive absorption of $CO_2$ is averagely 0.2 [34, 36–38].

### 2.5.2   Rate-Based Mode

The low values of the Murphree efficiency indicate that the process is far from the phase equilibrium condition. For this reason, the most used approach in the case of the reactive absorption-stripping of $CO_2$ with MEA is the so-called rate-based one [2, 8, 10, 13–21, 25–30, 39–45]. With this approach it is possible to take into account the limitations to mass transfer due to the presence of the chemical reactions. The rate-based mode was chosen for the description of both the absorber and the stripper and it is considered for the rest of the book. Furthermore, the rate-based

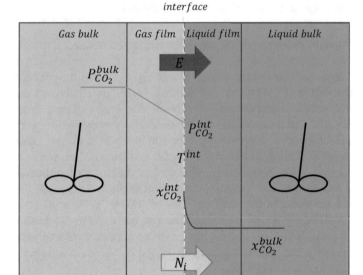

**Fig. 2.6** Rate-based segment representation for the absorption

model is based on the two-film theory of Lewis and Whitman [46]. A graphical representation of the rate-based segment for the absorption is reported in Fig. 2.6.

From Fig. 2.6 it can be noticed that the two phases, separated by the interface, are divided into two distinct zones: the bulk and the film. In the *RadFrac™* model—Rate-Based mode, different parameters and options must be set-up to describe the two zones. It must be highlighted that also the material and energy balances are different for the zones: the ideal CSTR balances are used for the bulk [47], while the rigorous Maxwell-Stefan approach is used for the interphase transfer and the material and energy transport in the films [48].

### 2.5.2.1 Bulk Modeling

The only input that must be set for what concerns the bulk is the *Flow Model*, which can be found in the *Packing Rating—Rate-Based modeling* panel in the *RadFrac™* block. Depending on the flow model, the bulk conditions are set equal to:

- the outlet conditions → *Mixed Flow*
- an average between the inlet and the outlet conditions → *Countercurrent*
- a combination between average and outlet conditions → *VPlug, VPlugP, LPlug*.

The *Mixed Flow* model was adopted in order to maintain the representation of the column as series of CSTRs. A comparison between the different flow models can be found in the works of Zhang et al. [14] and Errico et al. [18].

### 2.5.2.2  Film Modeling—Resistances

In Fig. 2.6, the profiles represented in the two films are related to the $CO_2$ molar fraction in the absorber. As it can be noticed, due to the absence of reactions in the gaseous phase, the profile is linear. After the $CO_2$ is absorbed into the liquid, it reacts very fast with the MEA [14, 18, 39]. This justifies the strongly nonlinear profile in the liquid film. To describe these different behaviors of the two films, different options are available in the *RadFrac™* model—Rate-Based mode to account for the specific resistances of each phase film. In particular, the option *Film* was used for the gaseous phase, since no reactions are present. In this case, only the resistance to material diffusion is considered. In the case of the liquid film, due to the presence of fast reactions, the film needs to be discretized and several parameters must be set-up in order to describe appropriately the steepness of the profiles. The option *Discrxn* was activated to consider contemporarily the resistances to diffusion, the presence of the reactions and to discretize the liquid film.

### 2.5.2.3  Film Modeling—Liquid Film Discretization

When the option *Discrxn* is set, the film is discretized into a certain number of points. The definition of specific parameters is crucial for the appropriate placement of the discretization points. From a physical point of view, since the reactions in the liquid film are fast, the profiles are expected to show a high steepness in the proximity of the interface. Then, it is preferable to use a small number of points placing them mostly close to the interface, as reported in the work of Kucka et al. [39], rather than use a high number of equidistant points. For this reason, the so-called *geometric discretization* was activated in the *Rate-Based Setup* panel. To make the discretization effective it is necessary to define three parameters:

- **Reaction Condition Factor (RCF)**. This parameter, which varies between 0 and 1, weights the interface and bulk compositions and temperature in the computation of the reaction rate within the film. When the reactions in the film are very fast, like in the case of the $CO_2$ capture with MEA, the bulk conditions must have a higher weight on the computation of the reaction rates. Then, a large value of the *RCF* must be set.
- **Number of discretization points in the film**.
- **Film Discretization Ratio (FDR)**. This parameter defines the segments width in the liquid film. In particular, for values larger than 1, the segments become smaller moving towards the interface. In this way, it is possible to concentrate

the discretization points next to interface. Anyway, there must be a compromise between the number of points and the *FDR*, since a too large value can lead to numerical problems due to the small discretization steps close to the interface.

### 2.5.2.4   Rate-Based Model Parameters Evaluation

An aspect that distinguishes the use of the rate-based approach from the equilibrium stage one is the need to evaluate several characteristic parameters. Once all the geometric features of the packing, i.e., type of packing, packing height and column diameter, are defined, a certain number of correlations must be chosen to evaluate the following rate-based model parameters:

- **Wetted surface area**. This parameter is defined as the wetted surface available per cubic meter of packing. In another words, it quantifies the area available for the exchange between the gaseous and the liquid phase;
- **Material and energy transfer coefficients**. These coefficients, defined as the ratio between the diffusion coefficients (material transfer) or the thermal conductivity (energy transfer) and the film thickness, are necessary to evaluate the interphase transfer flow rates.
- **Fractional liquid hold-up**. This parameter quantifies the percentage of free-volume occupied by the liquid. It is fundamental in the proper evaluation of the control volumes where material and energy balances are to be applied.

Depending on the packing type, i.e., structured or random, several correlations are available in the *RadFrac*™ block—*Packing Rating* panel for the computation of these parameters.

## 2.6   Analysis of the Fluid Dynamics

The specifications given in the previous sections were all related to the modeling of a single segment in the Rate-Based model. Another important aspect that must be considered is the one related to the system fluid dynamics. Typically, from a fluid dynamic point of view, reactive absorption-stripping columns are modeled as ideal plug-flow without axial dispersion [13, 25, 39, 42, 45]. In particular, the use of the *RadFrac*™ model implicates the approximation of the plug-flow as a series of n-CSTR, since in Aspen Plus® all the blocks contain algebraic equations. In order to correctly describe the fluid dynamics when modeling this kind of columns, it is necessary to verify if the column behavior resembles that of an ideal plug-flow. In particular, the possible presence of two phenomena can deviate the columns fluid dynamic behavior from the ideal one:

- The axial diffusion/dispersion;
- The backmixing due to the countercurrent.

It is then necessary to investigate if at least one of these two factors has an effect on the process.

### 2.6.1  Axial Diffusion/Dispersion—Peclet Number Analysis

The possible presence in the process of the axial diffusion/dispersion can be investigated by means of the computation of the Peclet number. This dimensionless group is defined as the ratio between the rate of transport by convection and the rate of transport by diffusion/dispersion, according Eq. 2.9:

$$Pe = \frac{FL_C}{\epsilon \psi SC\mathcal{D}} \tag{2.9}$$

where:

- $L_C$: characteristic length;
- $F$: molar flow rate;
- $\epsilon$: void fraction;
- $\psi$: fractional phase hold-up;
- $S$: column cross-sectional area;
- $C$: molar concentration;
- $\mathcal{D}$: diffusion coefficient.

In general, high values of the Peclet number indicate a column behavior close to an ideal plug-flow. On the other hand, if the value of the dimensionless group is small, the axial diffusion/dispersion has an effect on the process that cannot be neglected.

The Peclet number can be defined for both the phases either for what concerns the material transport and the energy transport. Moreover, two characteristic lengths can be considered, i.e., the column height and the packing equivalent diameter. The former provides information on the overall column behavior, while the latter considers the effect of the diffusion/dispersion at a local level, around the packing [18].

The expression of the Peclet number is obtained from the dimensionless form of the material and energy balances for a plug-flow with axial dispersion. This implicates that an appropriate reference condition must be selected to determine the value of the dimensionless number. The easiest choice is to consider the feed conditions (top for the liquid phase and bottom for the gaseous one) as reference for the physical properties. In this way, the evaluation of the Peclet number becomes independent from the system simulation.

## 2.6.2  Backmixing Due to the Countercurrent

The second factor which affects the ideal plug-flow behavior assumption is represented by the backmixing due to the countercurrent. This macroscale phenomenon is not taken into account by the Peclet number, which contains the axial diffusion/dispersion, a microscale phenomenon. The backmixing is implicitly included in the material and energy balances because of the countercurrent stream arrangement. To investigate its possible effect on the process it is necessary to obtain the correct numerical solution of the resulting system of equations.

## 2.6.3  The Number of Segments Analysis

The discussion on the modeling of the system fluid dynamics up to now dealt with the physics only, but it is strictly linked to a fundamental mathematical aspect which is the solution of the resulting system of equations obtained from the model development. Since in the *RadFrac*™ model the plug-flow is approximated as a series of CSTRs, a system of algebraic equations is solved by the software during the simulations. In this case, the number of CSTRs corresponds to the number of segments (or number of stages in Aspen Plus®) for the discretization of the axial domain in the packed columns. It is known from the theory that a series of CSTRs [47, 49] coincides with an ideal plug-flow when the number of CSTRs tends to infinite. Numerically speaking, this means that with a sufficiently high number of segments the series of CSTRs reasonably approximates an ideal plug-flow. Consequently, it is necessary to appropriately determine the number of discretization segments of the packing height for two main reasons:

- Investigate if the column behavior resembles that of an ideal plug-flow;
- Obtain the correct numerical solution of the system of equations.

Very often this parameter has been too easily defined in the literature, and its influence on the model has been overlooked. This is demonstrated by the fact that different research groups dealing with the same exact facility in the same operating conditions used a different number of segments for the discretization of the packing height [18]. To corroborate this trend, a list of some relevant works where the rate-based method was used to model a $CO_2$-MEA absorption system is reported in Table 2.1.

The analysis of Table 2.1 leads to the conclusion that there is not correspondence between the number of segments and the column dimensions. Even when the same column is considered, there is still an effective variation in the number of segments used. This disagreement indicates how the definition of this parameter is underestimated though, as it is going to be demonstrated in the next chapters, it is fundamental to obtain a correct model of the process.

**Table 2.1** Literature review on the number of segments applied in the rate-based model for the $CO_2$-MEA absorption system [19]

| Reference | Absorber | | |
|---|---|---|---|
| | Segments | Height (m) | Diameter (m) |
| [41] | 39 | 3.89 | 0.15 |
| [50] | 20 | 4.25 | 0.125 |
| [2] | 12 | 6.10 | 0.427 |
| [28] | 10 | 6.55 | 0.100 |
| [39] | 15 | 6.55 | 0.100 |
| [45] | 25 | 6.55 | 0.100 |
| [13] | 30 | 6.10 | 0.427 |
| [8] | 15 | 6.10 | 0.427 |
| [14] | 20 | 6.10 | 0.427 |
| [51] | 15 | 8.00 | 1.680 |
| [17] | 24 | 12.00 | 0.150 |
| [16] | 40 | 17.00 | 1.1 |

The definition of a correct number of segments for the discretization of the axial domain is strictly related to the possible influence that the axial diffusion/dispersion and the backmixing due to the countercurrent have on the process. In particular:

- For large values of the Peclet number, the axial diffusion/dispersion can be neglected. In this case the system fluid dynamics resembles that of an ideal plug-flow and, consequently, a high number of segments would be needed to correctly describe the process;
- Due to the countercurrent stream arrangement, even if the axial dispersion had no influence, the plug-flow assumption would fall if the backmixing generated by the countercurrent played an important role on the process. The investigation of the possible effect of the backmixing can be done only after the obtainment of the correct numerical solution of the resulting system of algebraic equations. This means that the number of segments must be increased until an asymptotic behavior is reached [52–54], i.e., until the differences in the profiles and the performance between two consecutive simulations with different number of segments are negligible. If the number of segments for the discretization of the axial domain necessary to obtain the correct solution is sufficiently high, it can be concluded that the backmixing due to the countercurrent has no important effect on the process.

The definition of a proper number of segments is of fundamental importance to obtain a representative model of the process. The application of the analysis of the system fluid dynamics is showed for both the absorber (Chap. 3) and the stripper (Chap. 4).

# References

1. Cau G, Tola V, Deiana P (2014) Comparative performance assessment of USC and IGCC power plants integrated with $CO_2$ capture systems. Fuel 116:820–833
2. Plaza JM, Wagener DV, Rochelle GT (2009) Modeling $CO_2$ capture with aqueous monoethanolamine. Energy Procedia 1(1):1171–1178
3. Wang M, Lawal P, Stephenson P et al (2011) Post-combustion $CO_2$ capture with chemical absorption: a state-of-the-art review. Chem Eng Res Des 89(9):1609–1624
4. Tan LS, Shariff M, Lau KK et al (2012) Factors affecting $CO_2$ absorption efficiency in packed column: a review. J Ind Eng Chem 18(6):1874–1883
5. Bui M, Gunawan I, Verheyen V et al (2014) Dynamic modelling and optimisation of flexible operation in post-combustion $CO_2$ capture plants-A review. Comput Chem Eng 61:245–265
6. Austgen DM, Rochelle GT, Peng X et al (1989) Model of Vapor-Liquid Equilibria for Aqueous Acid Gas-Alkanolamine Systems Using the Electrolyte-NRTL Equation. Ind Eng Chem Res 28(7):1060–1073
7. Weiland RH, Chakravarty T, Mather AE (1993) Solubility of carbon dioxide and hydrogen sulfide in aqueous alkanolamines. Ind Eng Chem Res 32(7):1419–1430
8. Lawal A, Wang M, Stephenson P et al (2009) Dynamic modelling of $CO_2$ absorption for post-combustion capture in coal-fired power plant. Fuel 88(12):2455–2462
9. Lin Y, Pan T-H, Shan-Hill Wong D et al (2011) Plantwide control of $CO_2$ capture by absorption and stripping using monoethanolamine solution. Ind Eng Chem Res 50(3):1338–1345
10. Biliyok C, Lawal A, Wang M et al (2012) Dynamic modelling, validation and analysis of post-combustion chemical absorption $CO_2$ capture plant. Int J Greenhouse Gas Control 9:428–445
11. Liu Y, Zhang L, Watanasiri S (1999) Representing vapor-liquid equilibrium for an aqueous MEA-$CO_2$ system using the electrolyte nonrandom-two-liquid model. Ind Eng Chem Res 38 (5):2080
12. Hilliard MD (2008) A predictive thermodynamic model for an aqueous blend of potassium carbonate, piperazine, and monoethanolamine for carbon dioxide capture from flue gas. Dissertation, The University of Texas at Austin
13. Kvamsdal HM, Rochelle GT (2008) Effect of the temperature bulge in $CO_2$ absorption from flue gas by aqueous monoethanolamine. Ind Eng Chem Res 47(3):867–875
14. Zhang Y, Chen H, Chen C-C et al (2009) Rate-based process modeling study of $CO_2$ capture with aqueous monoethanolamine solution. Ind Eng Chem Res 48(20):9233–9246
15. Moioli S, Pellegrini LA, Gamba S (2012) Simulation of $CO_2$ capture by MEA scrubbing with a rate-based model. Procedia Eng 42:1651–1661
16. Razi N, Svendsen HF, Bolland O (2013) Validation of mass transfer correlations for $CO_2$ absorption with MEA using pilot data. Int J Greenhouse Gas Control 19:478–491
17. Posch S, Haider M (2013) Dynamic modeling of $CO_2$ absorption from coal-fired power plant into an aqueous monoethanolamine solution. Chem Eng Res Des 91(6):977–987
18. Errico M, Madeddu C, Pinna D et al (2016) Model calibration for the carbon dioxide-amine absorption system. Appl Energy 183:958–968
19. Madeddu C, Errico M, Baratti R (2017) Rigorous modeling of a $CO_2$-MEA stripping system. Chem Eng Trans 57:451–456
20. Luo X, Wang M (2017) Improving prediction accuracy of a rate-based model of an MEA-based carbon capture process for large-scale commercial deployment. Engineering 3:232–243
21. Lawal A, Wang M, Stephenson P et al (2010) Dynamic modelling and analysis of post-combustion $CO_2$ chemical absorption process for coal-fired power plants. Fuel 89 (10):2791–2801
22. Freguia S (2002) Modeling of $CO_2$ Removal from Flue Gases with Monoethanolamine. Dissertation, The University of Texas at Austin

23. Nasrifar K, Tafazzol AH (2010) Vapor-liquid equilibria of acid gas-aqueous ethanolamine solutions using the PC-SAFT equation of state. Ind Eng Chem Res 49(16):7620–7630
24. Aboudheir A, Tontiwachwuthikul P, Chakma A et al (2003) Kinetics of the reactive absorption of carbon dioxide in high $CO_2$-loaded, concentrated aqueous monoethanolamine solutions. Chem Eng Sci 58(23–24):5195–5210
25. Tobiesen FA, Svendsen HF (2007) Experimental validation of a rigorous absorber model for $CO_2$ postcombustion capture. AIChE J 53(4):846–865
26. Faramarzi L, Kontogeorgis GM, Michelsen ML et al (2010) Absorber model for $CO_2$ capture by monoethanolamine. Ind Eng Chem Res 49(8):3751–3759
27. Meldon JH, Morales-Cabrera JA (2011) Analysis of carbon dioxide absorption in and stripping from aqueous monoethanolamine. Chem Eng Journ 171(3):753–759
28. Mores P, Scenna N, Mussati S (2012) A rate based model of a packed column for $CO_2$ absorption using aqueous monoethanolamine solution. Int J Greenhouse Gas Control 6:21–36
29. Mores P, Scenna N, Mussati S (2012) $CO_2$ capture using monoethanolamine (MEA) aqueous solution: Modeling and optimization of the solvent regeneration and $CO_2$ desorption process. Energy 45(1):1042–1058
30. Neveux T, Moullec YL, Corriou J-P et al (2013) Modeling $CO_2$ capture in amine solvents: prediction of performance and insights on limiting phenomena. Ind Eng Chem Res 52 (11):4266–4279
31. Aspen Technology, Inc. (2008) Aspen plus: rate based model of the $CO_2$ capture process by MEA using aspen plus. Aspen Technology Inc., Burlington, MA
32. Pinsent BRW, Pearson L, Roughton FJW (1956) The kinetics of combination of carbon dioxide with hydroxide ions. Trans Faraday Soc 52:1512–1520
33. Murphree EV (1925) Rectifying column calculations with particular reference to N component mixtures. Ind Eng Chem 17(7):747–750
34. Øi LE (2007) Aspen HYSYS simulation of $CO_2$ removal by amine absorption from a gas based power plant. Paper presented at the SIMS2007 Conference, Gøteborg, 30–31 October 2007
35. Mores P, Scenna N, Mussati S (2011) Post-combustion $CO_2$ capture process: Equilibrium stage mathematical model of the chemical absorption of $CO_2$ into monoethanolamine (MEA) aqueous solution. Chem Eng Res Des 89(9):1587–1599
36. Øi LE (2012) Comparison of Aspen HYSYS and Aspen Plus simulation of $CO_2$ absorption into MEA from atmospheric gas. Energy Procedia 23:360–369
37. Walter JF, Sherwood TK (1941) Gas absorption in bubble-cap columns. Ind Eng Chem 33 (4):493–501
38. Afkhamipour M, Mofarahi M (2013) Comparison of rate-based and equilibrium-stage models of a packed column for post-combustion $CO_2$ capture using 2-amino-2-methyl-1-propanol (AMP) solution. Int J Greenhouse Gas Control 15:186–199
39. Kucka L, Müller I, Kenig EY, Górak A (2003) On the modelling and simulation of sour gas absorption by aqueous amine solutions. Chem Eng Sci 58(16):3571–3578
40. Kvamsdal HM, Jakobsen JP, Hoff KA (2009) Dynamic modeling and simulation of a $CO_2$ absorber column for post-combustion $CO_2$ capture. Chem Eng Process Intensif 49(1):135–144
41. Gáspár J, Cormoş A-M (2011) Dynamic modeling and validation of absorber and desorber column for post-combustion $CO_2$ capture. Comput Chem Eng 35(10):2044–2052
42. Khan FM, Krishnamoorti V, Mahmud T (2011) Modelling reactive absorption of $CO_2$ in packed column for post-combustion carbon capture applications. Chem Eng Res Des 89 (9):1600–1608
43. Gaspar J, Cormos A-M (2012) Dynamic modeling and absorption capacity assessment of $CO_2$ capture process. Int J Greenhouse Gas Control 8:45–55
44. Kvamsdal HM, Hillestad M (2012) Selection of model parameter correlations in a rate-based $CO_2$ absorber model aimed for process simulation. Int J Greenhouse Gas Control 11:11–20

45. Mac Dowell N, Samsatli NJ, Shah N (2013) Dynamic modelling and analysis of an amine-based post-combustion $CO_2$ capture absorption column. Int J Greenhouse Gas Control 12:247–258
46. Lewis WK, Whitman WG (1924) Principles of gas absorption. Ind Eng Chem 16(12):1215–1220
47. Scott-Fogler H (2006) Elements of chemical reaction engineering, Prentice Hall
48. Taylor R, Krishna R (1993) Multicomponent Mass Transfer. Wiley, New York
49. Levenspiel O (1999) Chemical reaction engineering. Wiley, New York
50. Zhang Y, Chen C-C (2013) Modeling $CO_2$ absorption and desorption by aqueous monoethanolamine solution with Aspen rate-based model. Energy Procedia 37:1584–1596
51. Pacheco MA, Rochelle GT (1998) Rate-based modeling of reactive absorption of $CO_2$ and $H_2S$ into aqueous methyldiethanolamine. Ind Eng Chem Res 37(10):4107–4117
52. Davis ME (1984) Numerical methods and modelling for chemical engineers. Wiley, New York
53. Kenig EY, Schneider R, Górak A (1999) Rigorous dynamic modelling of complex reactive absorption processes. Chem Eng Sci 54(21):5195–5203
54. Schneider R, Kenig EY, Górak A (2001) Complex reactive absorption processes: model optimization and dynamic column simulation. Comput Aided Chem Eng 9:285–290

# Chapter 3
# Model Validation for the Absorber

In this chapter, the absorption section of two pilot-plant facilities, different in dimensions and operating conditions, is modeled. A discussion on the typical temperature bulge, which gives important indications on the behavior of the absorber, is reported. The Peclet number is evaluated for both the cases with reference to the material and the energy transport and to different characteristic lengths, highlighting that the columns have a plug-flow like behavior. Then, the number of segments analysis is performed in order to obtain the correct representation of the process and investigate the possible effect of the backmixing due to the countercurrent. The proposed model suitably describes the experimental data and particularly the temperature bulge, independently on its location.

## 3.1 Absorption Section Case Studies

The model developed in Chap. 2 was first validated for the absorber. Two experimental pilot-plant facilities, different in dimensions and operating conditions were chosen to test the model:

- the laboratory-scale absorption plant designed in the work of Tontiwachwuthikul et al. [1, 2];
- the large-scale absorption/desorption system CESAR (CO$_2$ Enhanced Separation and Recovery) described by Razi et al. [3].

The two systems were chosen because of their differences in dimensions (laboratory- vs. large-scale) and type of packing (random vs. structured).

In both facilities, the absorption system consists in a packed column divided into a certain number of sections separated by redistributors. The main features of the packing and the column dimensions are reported in Table 3.1.

For each plant, one run was selected for the model validation. In particular, Run T22 was chosen from the lab-scale facility, while Run 1-A2 was considered from

© The Author(s), under exclusive license to Springer Nature Switzerland AG 2019
C. Madeddu et al., *CO$_2$ Capture by Reactive Absorption-Stripping*,
SpringerBriefs in Energy, https://doi.org/10.1007/978-3-030-04579-1_3

**Table 3.1** Column and packing features for the two facilities

| Source | Tontiwachwuthikul et al. [2] | Razi et al. [3] |
|---|---|---|
| Packing height (m) | 6.55 | 17 |
| Column diameter (m) | 0.1 | 1.1 |
| Packing type | 12.7 mm Berl Saddles | Sulzer Mellapak 2X |
| Void fraction ($m^3/m^3$) | 0.62 | 0.99 |
| Dry specific area ($m^2/m^3$) | 465 | 205 |

**Table 3.2** Feed characterization for the two selected runs

| Source | Tontiwachwuthikul et al. [2] | | Razi et al. [3] | |
|---|---|---|---|---|
| Run | T22 | | 1-A2 | |
| Stream | Flue gas | Lean amine | Flue gas | Lean amine |
| Temperature (K) | 288.15 | 292.15 | 326.92 | 332.57 |
| Molar flow (mol/s) | 0.14 | 1.04 | 52.33 | 214.55 |
| $CO_2$ (mol frac) | 0.191 | 0 | 0.12 | 0.0263 |
| MEA (mol frac) | 0 | 0.055 | 0 | 0.102 |
| $H_2O$ (mol frac) | 0.1 | 0.945 | 0.12 | 0.8717 |
| $N_2$ (mol frac) | 0.709 | 0 | 0.76 | 0 |
| Pressure (kPa) | 103.15 | 103.15 | 106.391 | 101.325 |

**Table 3.3** Experimental data for Run T22 from the lab-scale plant

| Run | | | T22 | |
|---|---|---|---|---|
| Sample | $H$ (m) | | $T^L$ (K) | $y_{CO_2}$ (mol frac) |
| 1 | 0.00 | | 321.15 | 0.191 |
| 2 | 1.05 | | 330.15 | 0.128 |
| 3 | 2.15 | | 320.15 | 0.053 |
| 4 | 3.25 | | 305.15 | 0.012 |
| 5 | 4.35 | | 295.15 | 0.001 |
| 6 | 5.45 | | 293.15 | 0.000 |
| 7 | 6.55 | | 292.15 | 0.000 |

the large-scale plant. The feed characterization, i.e., the input values for the streams in Aspen Plus®, are reported for both the runs in Table 3.2.

A certain number of experimental data is available for each plant. In the lab-scale plant, each section is equipped with sensors for the measurement of the liquid temperature and the $CO_2$ vapor composition in the gaseous phase. On the other hand, a sample point for the measurement of the temperature is present in the large-scale plant for each section. Furthermore, for both plants, the experimental absorber performance values are available. The list of the experimental data is reported in Tables 3.3 and 3.4 for Run T22 and Run 1-A2, respectively.

**Table 3.4** Experimental data for Run 1-A2 from the large-scale plant

| Run | | 1-A2 |
|---|---|---|
| Sample | $H$ (m) | $T^L$ (K) |
| 1 | 0.00 | 326.92 |
| 2 | 4.25 | 333.47 |
| 3 | 8.50 | 339.95 |
| 4 | 12.75 | 346.55 |
| 5 | 17 | 332.57 |

## 3.2 The Temperature Bulge

The $CO_2$ reactive absorption process is characterized by several heat transfer phenomena, including those related to the reactions (which are exothermic in the case of the absorption) and water vaporization/condensation. For this reason, the absorber temperature profiles typically show a pronounced bulge. The position and magnitude of this bulge give important indications on the behavior of the process which have fundamental implications in the design of the absorber, as it is going to be demonstrated in Chap. 5. The features of the bulge are influenced by different factors, as extensively studied in the work of Kvamsdal and Rochelle [4]. In particular, the ratio between the liquid and the gas molar flow rates, i.e., the Liquid to Vapor ratio, L/V, is one of the main factors. Typically, three situations are possible depending on the value of this parameter:

- L/V < 5: the bulge is located at the top of the column;
- 5 < L/V < 6: the bulge is located in the middle of the column;
- L/V > 6: the bulge is located at the bottom of the column.

The first situation is usually verified in columns filled with structured packing, since due to the higher surface area, the required performance can be achieved with a low L/V ratio value. On the other hand, when columns filled with random packing are considered, a higher L/V ratio is required to obtain the requested separation performance. The two runs selected were chosen according to the possibility of different temperature bulge positions. In particular, in the random packed lab-scale plant, the L/V ratio is equal to 7.43 for Run T22 and, consequently, a bottom temperature bulge is expected. On the other hand, a 4.1 value for the L/V ratio was found in Run 1-A2 and a top temperature bulge was awaited for the large-scale absorber, which is filled with structured packing [5].

## 3.3 Peclet Number Analysis

Before the simulations, the Peclet number analysis was performed for the two runs. It was outlined in Sect. 2.6.1 how the evaluation of the dimensionless group gives indications on the influence of the axial diffusion/dispersion on the process, which

**Table 3.5** Peclet number evaluation for Run T22

| Run | T22 | |
|---|---|---|
| Characteristic length | $H$ | $d_{eq}$ |
| $Pe^L_{M,mix}$ | $2.09 \times 10^9$ | $1.57 \times 10^6$ |
| $Pe^G_{M,mix}$ | $2.75 \times 10^5$ | 205.93 |
| $Pe^L_T$ | $3.16 \times 10^6$ | 2364.74 |
| $Pe^G_T$ | $2.92 \times 10^5$ | $2.19 \times 10^2$ |

**Table 3.6** Peclet number evaluation for Run 1-A2

| Run | 1-A2 | |
|---|---|---|
| Characteristic length | $H$ | $d_{eq}$ |
| $Pe^L_{M,mix}$ | $3.12 \times 10^9$ | $3.9 \times 10^6$ |
| $Pe^G_{M,mix}$ | $1.1 \times 10^6$ | 1298.1 |
| $Pe^L_T$ | $2.03 \times 10^7$ | $2.39 \times 10^4$ |
| $Pe^G_T$ | $1.19 \times 10^6$ | $1.40 \times 10^3$ |

is a fundamental factor in the choice of a proper number of segments for the discretization of the axial domain. The Peclet number was evaluated for both the material (for the mixture) and energy transport in each phase and with reference to both the packing height and the packing equivalent diameter using Aspen Custom Modeler®. The inlet conditions (column top for the liquid phase, column bottom for the gaseous phase) were used as reference conditions. The values of the Peclet number for the liquid and the gaseous mixture are reported in Tables 3.5 and 3.6 for Run T22 and Run 1-A2, respectively.

As it is possible to notice from the Peclet number analysis the values are quite large. In fact, the minimum values, which are observed for the gas phase in all the cases when the packing equivalent diameter is considered as characteristic length, are in the order of $10^2$ for Run T22 and $10^3$ for Run 1-A2. This difference was somehow expected, due to the differences in dimensions and flow rates involved in the two facilities. From these results, it can be concluded that the axial diffusion/dispersion has no effect on the process and that the columns fluid dynamics resembles that of an ideal plug-flow [5]. Then, from a mathematical point of view, a sufficiently high number of segments for the discretization of the axial domain is required to obtain the correct solution of the resulting system of algebraic equations.

## 3.4   Backmixing Due to the Countercurrent Effect

The results obtained in the previous section with the Peclet number analysis led to the conclusion that the axial diffusion/dispersion could be neglected and, consequently, the columns behavior from a fluid dynamic point of view resembles that of an ideal plug-flow. However, in the reality, the countercurrent generates a backmixing effect that can deviate the real columns behavior form the ideal plug-flow one. Being a macroscale phenomenon, the backmixing is not taken into account in the Peclet number analysis, since the dimensionless group contains the axial diffusion/dispersion, which is a microscale phenomenon. The backmixing due to the countercurrent is implicitly present in the material and energy balances describing the process and its possible effect can be analyzed only after the obtainment of the correct numerical solution of the system. In particular, if a high number of segments is needed to find the correct solution, then it can be concluded that the backmixing does not play and important role in the process.

Finally, it is worth to remember that, using the Aspen Plus® *RadFrac*™ model—Rate-Based mode, the plug-flow is approximated as a series of n-CSTRs, where the number of CSTRs corresponds to the number of segments for the discretization of the axial domain.

Keeping constant the input parameters reported in Tables 3.1 and 3.2, the number of segments was varied until the difference between two consecutive sets of profiles became negligible [6–8].

## 3.5 Rate-Based Model Set-Up

### 3.5.1 Rate-Based Correlations

As previously reported in Sect. 2.5.2.3, in the case of the rate-based model several parameters must be evaluated to characterize the interphase transfer and the control volumes for the balances. Depending on the kind of packing, a list of correlations is present in the *Packing Rating—Rate-Based* section of the *RadFrac*™ block. Table 3.7 resumes the correlations used for the lab-scale and the large-scale plants.

It must be highlighted that in the case of the random packing, the correlation by Onda et al. [9] underestimates the wetted surface area, as reported in the work of Zhang et al. [14]. For this reason, the *Interfacial Area Factor*, available in the simulator to correct the evaluation of the wetted surface area, was set to 1.2. This value is coherent with the dry specific area for the ceramic Berl saddles reported by Mores et al. [15].

**Table 3.7** Rate-based correlations used for the two runs

| Source | Tontiwachwuthikul et al. [2] | Razi et al. [3] |
|---|---|---|
| Run | T22 | 1-A2 |
| Packing type | Random | Structured |
| Wetted surface area | Onda et al. [9] | Bravo et al. [10] |
| Material transfer coefficients | Onda et al. [9] | Bravo et al. [10] |
| Heat transfer coefficients | Chilton and Colburn [11] | Chilton and Colburn [11] |
| Fractional liquid hold-up | Stichlmair et al. [12] | Bravo et al. [13] |

### 3.5.2  Liquid Film Discretization Parameters

Due to activation of the option *Discrxn* in the *Packing Rating—Rate-Based* panel, the three parameters reported in Sect. 2.5.2.3 must be defined in the *Rate-Based Modeling—Rate-Based Setup* section of the software to make the discretization of the liquid film effective:

- the *Reaction Condition Factor* was set to 0.9 in order to give more weight to the bulk conditions in the evaluation of the film reaction rates, due to the fast reactions in the liquid film. This value is in agreement with the work of Zhang et al. [14];
- after a significant number of simulations, it was found that 5 non-equidistant *Discretization Points* in the liquid film were sufficient for the correct description of the profiles on the basis of the comparison between the model results and the experimental data. This value is in agreement with the work of Kucka et al. [16];
- the *Film Discretization Ratio* was fixed to 10, value that was found to be a good compromise between the discretization points placement and the discretization steps for the numerical solution of the system.

## 3.6    Laboratory-Scale Plant: Run T22

### 3.6.1  Number of Segments Analysis

Run T22 from the lab-scale plant is the first analyzed. Following the procedure illustrated in Sect. 2.6.3, the plant was simulated varying the number of segments until two consecutive profiles were overlapped. The profiles for the liquid temperature and the $CO_2$ composition in the vapor phase for different values of the number of segments are reported in Fig. 3.1

From the analysis of Fig. 3.1 it is possible to notice that the profiles with 80 and 90 segments are overlapped. Moreover, Fig. 3.1a shows that the temperature bulge, located in this case at the bottom of the absorber, is correctly described only after at least 30 segments. On the other hand, Fig. 3.1b highlights that only with an appropriate number of segments it is possible to identify a mild concentration bulge in the bottom of the column, which is caused by the partial preponderance of the carbamate ion dissociation back to $CO_2$ and MEA. This result confirms the indication given by the Peclet number analysis and leads also to the conclusion that the backmixing effect can be neglected. The number of segments analysis was needed to obtain a correct process model from a numerical point of view.

For what concerns the variation of the column performance evaluation with the number of segments, the values are reported in Table 3.8. For Run T22, the performance values are represented by the $CO_2$ removal percentage and the rich solvent loading.

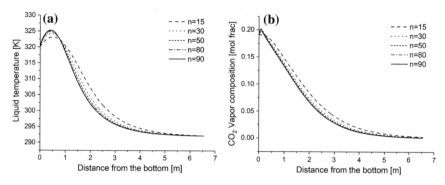

**Fig. 3.1** Absorber **a** liquid temperature and **b** $CO_2$ vapor composition profiles variation with the number of segments

**Table 3.8** Absorber performance variation with the number of segments for Run T22

| Performance | Number of segments | | | | |
|---|---|---|---|---|---|
| | 15 | 30 | 50 | 80 | 90 |
| $CO_2$ removal % | 99.12 | 99.45 | 99.56 | 99.62 | 99.63 |
| Loading out | 0.455 | 0.457 | 0.457 | 0.457 | 0.457 |

**Fig. 3.2** Absorber $CO_2$ interphase molar flow rate profile variation with the number of segments

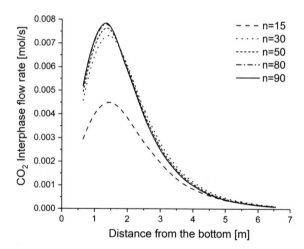

It is possible to observe that the values of both the $CO_2$ removal and the loading in the rich solvent grow with the increase of the number of segments. This trend can be explained analyzing the variation of the $CO_2$ interphase molar flow rate profiles with the number of segments, reported in Fig. 3.2.

From the analysis of Fig. 3.2 it can be noticed that when the number of segments is increased the simulator evaluates a higher mean $CO_2$ transfer flow rate from the gaseous to the liquid phase due to the more detailed discretization of the axial domain.

**Fig. 3.3** Absorber $H_2O$ interphase molar flow rate profile variation with the number of segments

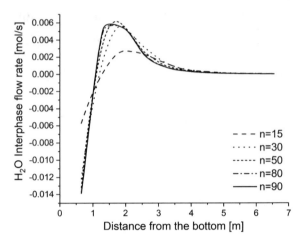

These first results highlight the need of a sufficiently high number of segments to discretize the axial domain. This necessity could be explained also by the fact that along the column several transitions happen between the absorption and the desorption process regimes and between water evaporation and condensation [7]. Using an inadequate number of segments, the net fluxes, especially for water, between liquid and vapor could be under/over-estimated, as it is showed in Fig. 3.3, leading to different temperature and composition profiles. It must be specified that in Fig. 3.3 a *negative* flow rate indicates a transfer from the liquid to the gaseous phase (evaporation), while a *positive* flow rate indicates a transfer from the gaseous to the liquid phase (condensation).

### 3.6.2   Comparison with the Experimental Data

Once the appropriate number of segments is set, it is possible to make the comparison between the model and the experimental data, reported in Fig. 3.4.

A good agreement between the model with 90 segments and the experimental values is obtained for both the liquid temperature and the $CO_2$ composition in the gaseous phase. This result is corroborated by the evaluation of the standard error (*SE*), defined as the square root of the mean squared error (*MSE*), as reported in Eq. 3.1:

$$SE = \sqrt{MSE} \tag{3.1}$$

For Run T22, the *SE* related to the liquid temperature was 0.74 K, while $5 \times 10^{-3}$ was found for the $CO_2$ molar fraction in the gaseous phase.

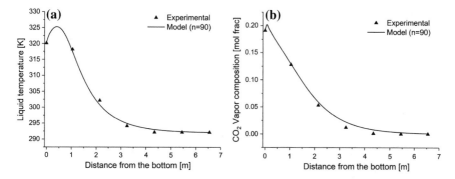

**Fig. 3.4** Comparison between absorber **a** liquid temperature and **b** $CO_2$ vapor composition profiles with the experimental data

**Table 3.9** Comparison between the experimental performance and the model for Run T22

| Performance | Experimental | Model (n = 90) |
|---|---|---|
| $CO_2$ removal % | 100 | 99.63 |
| Loading out | 0.443 | 0.457 |

Finally, the comparison between the model and the experimental values of the absorber performance are reported in Table 3.9. Again, the results show a good agreement between the model and the experimental data.

## 3.7 Large-Scale Plant: Run 1-A2

In order to test the ability of the proposed model to describe different sets of experimental data independently of the column dimensions and the operating conditions, Run 1-A2 from the large-scale plant was simulated following the same procedure used for Run T22 of the lab-scale plant. In this case, the column is filled with structured packing and the flow rates involved are higher compared to the previous case. Only temperature measurements were available for this plant and the phase was not specified. Then, Fig. 3.5 reports the liquid and the vapor/gas temperature profiles variation with the number of segments.

In this case, 140 segments are needed to obtain the correct mathematical solution of the system of algebraic equations. This higher value is justified by the higher values of the Peclet number for the large-scale plant. The temperature bulge, due to the L/V ratio less than 5, appears in the top of the column.

Once the appropriate number of segments was determined, the comparison between the model and the experimental data was reported in Fig. 3.6.

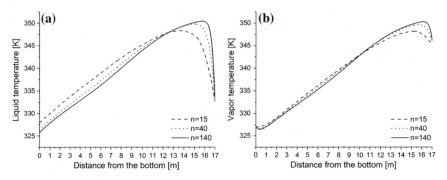

**Fig. 3.5**  Absorber **a** liquid temperature and **b** vapor temperature profiles variation with the number of segments

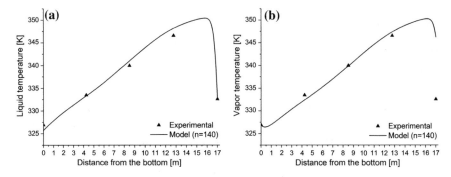

**Fig. 3.6**  Comparison between absorber **a** liquid temperature and **b** vapor temperature profiles with the experimental data

Again, a good agreement is found between the model and the experimental data. For what concerns the column performance, the model with 140 segments predicts a $CO_2$ removal of 87.2% while the experimental value was 90%.

In the light of these results, it was proved the ability of the model to describe the experimental data even in the case when the temperature bulge is located at the top of the absorber. Moreover, it was observed again the necessity to use an appropriate number of segments to simulate the process correctly.

The use of a too low number of segments leads to an incorrect description of the critical temperature bulge zone and to an incorrect evaluation of the internal fluxes.

# References

1. Tontiwachwuthikul P, Meisen A, Lim CJ (1989) Novel pilot plant technique for sizing gas absorbers with chemical reactions. Can J Chem Eng 67(4):602–607
2. Tontiwachwuthikul P, Meisen A, Lim CJ (1992) $CO_2$ absorption by NaOH, Monoethanolamine and 2-Amino-2-Methyl-1-Propanol solutions in a packed column. Chem Eng Sci 47(2):381–390
3. Razi N, Svendsen HF, Bolland O (2013) Validation of mass transfer correlations for $CO_2$ absorption with MEA using pilot data. Int J Greenhouse Gas Control 19:478–491
4. Kvamsdal HM, Rochelle GT (2008) Effect of the temperature bulge in $CO_2$ absorption from flue gas by aqueous monoethanolamine. Ind Eng Chem Res 47(3):867–875
5. Errico M, Madeddu C, Pinna D et al (2016) Model calibration for the carbon dioxide-amine absorption system. Appl Energy 183:958–968
6. Davis ME (1984) Numerical methods and modelling for chemical engineers. Wiley, New York
7. Kenig EY, Schneider R, Górak A (1999) Rigorous dynamic modelling of complex reactive absorption processes. Chem Eng Sci 54(21):5195–5203
8. Schneider R, Kenig EY, Górak A (2001) Complex reactive absorption processes: model optimization and dynamic column simulation. Comput Aided Chem Eng 9:285–290
9. Onda K, Takeuchi H, Okumoto Y (1968) Mass transfer coefficients between gas and liquid phases in packed columns. J Chem Eng Jpn 1(1):56–62
10. Bravo JL, Ja Rocka, Fair JR (1985) Mass transfer in gauze packings. Hydrocarb Processes 64 (1):91–95
11. Chilton TH, Colburn AP (1934) Mass transfer (absorption) coefficient. Prediction from data on heat transfer and fluid friction. Ind Eng Chem 26(11):1183–1187
12. Stichlmair J, Bravo JL, Fair JR (1989) General model for prediction of pressure drop and capacity of countercurrent gas/liquid packed columns. Gas Sep Purif 3(1):19–28
13. Bravo JL, Rocha JA, Fair JR (1992) A comprehensive model for the performance of column containing structured packing. IChEME Symp S 128:A439–A453
14. Zhang Y, Chen H, Chen C-C et al (2009) Rate-based process modeling study of $CO_2$ capture with aqueous monoethanolamine solution. Ind Eng Chem Res 48(20):9233–9246
15. Mores P, Scenna N, Mussati S (2012) A rate based model of a packed column for $CO_2$ absorption using aqueous monoethanolamine solution. Int J Greenhouse Gas Control 6:21–36
16. Kucka L, Müller I, Kenig EY, Górak A (2003) On the modelling and simulation of sour gas absorption by aqueous amine solutions. Chem Eng Sci 58(16):3571–3578

# Chapter 4
# Model Validation for the Stripper

In this chapter, the stripping section is modeled and experimental data from two different pilot-plant facilities are used for the validation. In Aspen Plus®, two different sets of degrees of freedom are chosen for the two plants, in order to study the effect of the proposed procedure on the evaluation of the output streams features in one case and on the evaluation of the reboiler duty in the other case. After the Peclet number evaluation, the number of segments analysis is performed. The obtained model is able to describe correctly the experimental data, validating the proposed procedure for the stripper. Moreover, it is highlighted how a correct use of the software leads to a better estimation of the reboiler duty.

## 4.1 Introduction to the Stripping Section Modeling

After the model validation for the absorber described in Chap. 3, this chapter is dedicated to the validation of the model for what concerns the stripping section. The stripper represents undoubtedly the most critical part of the system from an economical point of view, and its optimization is crucial in order to minimize the energy consumption in the capture plant. Since the reboiler duty represents by far the highest operating cost of the plant, as it drives the entire thermal swing, it is also the most important variable for the whole plant improvement.

The development of a model able to correctly describe the behavior of the stripper is an essential step to successively identify new possible ways to minimize the energy consumption. As highlighted by Tobiesen et al. [1], the solvent regeneration section is more complex than the absorption one due to the presence of condenser and reboiler. For this reason, a detailed model of this section is needed to have a better description of the phenomena happening inside the equipment and to improve the interpretation of the results obtained from an experimental campaign.

© The Author(s), under exclusive license to Springer Nature Switzerland AG 2019  43
C. Madeddu et al., *CO2 Capture by Reactive Absorption-Stripping*,
SpringerBriefs in Energy, https://doi.org/10.1007/978-3-030-04579-1_4

## 4.2   Stripping Section Case Studies

Similarly to the case of the absorber, two different pilot plant facilities were considered to validate the model developed in Chap. 2:

- the $CO_2$ capture facility from SINTEF Material and Chemistry described in the work of Tobiesen et al. [1];
- the pilot plant from the University of Texas at Austin (UTA) described in the work of Dugas [2].

In both plants, the stripping section contains a packed column with partial condenser and reboiler. One important difference between the two cases is found in the configuration for the water stream exiting the condenser: while in the SINTEF case the water is mixed with the liquid exiting the bottom of the stripper and then sent to the reboiler, in the UTA case the same stream is sent back to the top of the column as reflux. The main features of the packing and the column dimensions are reported in Table 4.1.

For each plant, one run was selected for the model validation. In particular, Run 1 was considered from the SINTEF plant, while Run 47 was chosen from the UTA plant. The feed characterization and the operating conditions are reported for both the runs in Tables 4.2 and 4.3, respectively.

**Table 4.1**   Column and packing features for the two facilities

| Source | Tobiesen et al. [1] | Dugas [2] |
|---|---|---|
| Packing height (m) | 3.89 | 6.1 |
| Column diameter (m) | 0.1 | 0.427 |
| Packing type | Sulzer Mellapak 250Y | Flexipac 1Y |
| Void fraction $(m^3/m^3)$ | 0.987 | 0.91 |
| Dry specific area $(m^2/m^3)$ | 256 | 420 |

**Table 4.2**   Feed characterization for the two selected runs

| Source | Tobiesen et al. [1] | Dugas [2] |
|---|---|---|
| Run | 1 | 47 |
| Stream | Rich amine | Rich amine |
| Temperature (K) | 389.91 | 356 |
| Molar flow (kmol/h) | 10.71 | 90.99 |
| $CO_2$ (mol frac) | 0.0348 | 0.0534 |
| MEA (mol frac) | 0.1102 | 0.1181 |
| $H_2O$ (mol frac) | 0.8549 | 0.828 |
| $N_2$ (mol frac) | 0 | 0.0005 |
| Pressure (kPa) | 196.96 | 196.96 |

**Table 4.3** Operating conditions for the two selected runs

| Source | Tobiesen et al. [1] | Dugas [2] |
|---|---|---|
| Run | 1 | 47 |
| Condenser/Top pressure (kPa) | 196.96 | 68.95 |
| Pressure drop (kPa) | 1 | 0.41 |
| Condenser temperature (K) | 288.15 | 297 |
| Reboiler duty (kW) | 11.6 | 205 |

For both plants, experimental liquid temperature values along the column are available. The list of the experimental data is reported in Tables 4.4 and 4.5 for Run 1 and Run 47, respectively.

Furthermore, different experimental values for the output streams are also available depending on the plant, as resumed in Tables 4.6 and 4.7 for Run 1 and Run 47, respectively.

**Table 4.4** Experimental data for Run 1 from the SINTEF plant

| Run | | | 1 |
|---|---|---|---|
| Sample | | $H$ (m) | $T^L$ (K) |
| 1 | | 0.01 | 393.59 |
| 2 | | 1.05 | 392.41 |
| 3 | | 2.15 | 391.5 |
| 4 | | 3.25 | 391 |
| 5 | | 4.35 | 389.81 |

**Table 4.5** Experimental data for Run 47 from the UTA plant

| Run | | | 47 |
|---|---|---|---|
| Sample | | $H$ (m) | $T^L$ (K) |
| 1 | | 1.75 | 365.16 |
| 2 | | 3.05 | 363.98 |
| 3 | | 4.72 | 363.47 |
| 4 | | 6.245 | 361.16 |

**Table 4.6** Output streams experimental data for Run 1 from the SINTEF plant

| Run | 1 |
|---|---|
| *Lean solvent* | |
| Temperature (K) | 394.15 |
| $CO_2$ (mol frac) | 0.0237 |
| MEA (mol frac) | 0.1086 |
| $H_2O$ (mol frac) | 0.8677 |
| Loading (mol $CO_2$/mol MEA) | 0.2185 |
| *Gas from condenser* | |
| Molar flow (kmol/h) | 0.1159 |

| Run | 47 |
|---|---|
| *Lean solvent* | |
| Loading (mol $CO_2$/mol MEA) | 0.28 |
| *Gas from condenser* | |
| $CO_2$ molar flow (kmol/h) | 92 |

**Table 4.7** Output streams experimental data for Run 47 from the UTA plant

## 4.3   Stripper Degrees of Freedom

Differently from the absorber, due to the presence of the condenser and the reboiler, two degrees of freedom must be defined for the stripper in order to perform the simulations. Two different sets of degrees of freedom were chosen for the two plants studied. In particular:

- in the case of the SINTEF plant, the condenser temperature and the reboiler duty were fixed;
- in the case of the UTA plant, the condenser temperature and the $CO_2$ gas molar flow rate were fixed.

This choice of the degrees of freedom was done to prove the validity of the proposed model in the case of different experimental data sets available. Moreover, as it is going to be showed in the next sections, this kind of analysis proves once again the important of having a correct model of the process, with outcomes that have important implications both at design and dynamic level.

## 4.4   Peclet Number Analysis

The first step in the procedure regards the Peclet number analysis, which was performed for all the runs examined. Like the absorber, the Peclet number was evaluated for both the material (for the mixture) and energy transport in each phase and with reference to both the packing height and the packing equivalent diameter using Aspen Custom Modeler®. The inlet conditions (column top for the liquid phase, column bottom for the gaseous phase) were used as reference conditions. The values of the Peclet number for the liquid and the gaseous mixture are reported in Tables 4.8 and 4.9 for Run 1 and Run 47, respectively.

**Table 4.8** Peclet number evaluation for Run 1

| Run | 1 | |
|---|---|---|
| Characteristic length | $H$ | $d_{eq}$ |
| $Pe^L_{M,mix}$ | $2.00 \times 10^8$ | $6.55 \times 10^5$ |
| $Pe^G_{M,mix}$ | $1.38 \times 10^5$ | 453.98 |
| $Pe^L_T$ | $4.22 \times 10^6$ | $1.38 \times 10^4$ |
| $Pe^G_T$ | $1.69 \times 10^5$ | 554.6 |

**Table 4.9** Peclet number evaluation for Run 47

| Run | 47 | |
|---|---|---|
| Characteristic length | $H$ | $d_{eq}$ |
| $Pe^L_{M,mix}$ | $3.96 \times 10^7$ | $5.85 \times 10^4$ |
| $Pe^G_{M,mix}$ | $4.36 \times 10^5$ | 643.57 |
| $Pe^L_T$ | $5.56 \times 10^5$ | 819.71 |
| $Pe^G_T$ | $4.92 \times 10^5$ | 725.96 |

From the analysis of Tables 4.8 and 4.9, it can be observed that even for the stripper the values of the Peclet number are quite large. Like the absorber, since the lowest value of the dimensionless group is in the order of $10^2$, it can be concluded that the axial diffusion/dispersion has no effect on the process and that the column fluid dynamics resembles that of an ideal plug-flow [3, 4]. Consequently, even in the case of the stripper, a sufficiently high number of segments is needed to obtain the correct solution of the resulting system of algebraic equations.

## 4.5 SINTEF Plant: Run 1

After the evaluation of the Peclet number, the SINTEF plant was simulated varying the number of segments until two consecutive profiles were overlapped to obtain the correct solution of the system from a numerical point of view. It must be remembered that in this case the condenser temperature and the reboiler duty are fixed to saturate the stripper degrees of freedom. Since the water from the condenser is not sent to the top of the column as reflux but it is mixed with the liquid from the stripper bottom, both condenser and reboiler were modeled separately from the column using the *Flash2* model in Aspen Plus®. For this reason, it was possible to fix the degrees of freedom directly in the software. It is worth to specify that the parameters for the set-up of the rate-based model are the same used in Chap. 3 for the absorber. The variation of the liquid temperature profile with the number of segments is reported in Fig. 4.1

As it is possible to notice from the analysis of Fig. 4.1, the asymptotic behavior is reached with 70 segments.

After the evaluation of a proper number of segments it is possible to make the comparison between the model and the experimental data, reported in Fig. 4.2.

The good agreement between the model and the experimental data is demonstrated by the evaluation of the standard error, defined in Eq. 3.1, which is 0.23 K in this case. The same good agreement is observed for what concerns the output measured variables, as reported in Table 4.10.

In the case of the stripper, differently from the absorber, since no significant gradients are present, the differences between the model with 10 segments and the model with 70 segments might not seem evident from the analysis of the liquid temperature profiles reported in Fig. 4.1. However, considering the variation of the

**Fig. 4.1** Stripper liquid temperature profile variation with the number of segments for Run 1

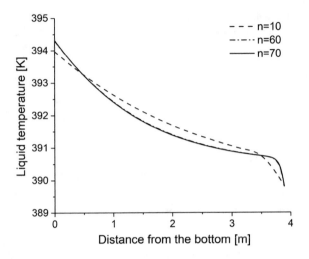

**Fig. 4.2** Comparison between the model liquid temperature profile and the experimental data

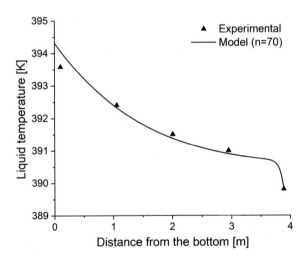

**Table 4.10** Comparison between the output streams experimental data and the model results

| Variable | Experimental | Model (n = 70) |
|---|---|---|
| *Lean solvent* | | |
| Temperature (K) | 394.15 | 395.24 |
| $CO_2$ (mol frac) | 0.0237 | 0.026 |
| MEA (mol frac) | 0.1086 | 0.1113 |
| $H_2O$ (mol frac) | 0.8677 | 0.8627 |
| Loading (mol $CO_2$/mol MEA) | 0.2185 | 0.2338 |
| *Gas from condenser* | | |
| Molar flow (kmol/h) | 0.1159 | 0.0981 |

$CO_2$ vapor molar fraction with the number of segments, reported in Fig. 4.3, it can be observed that the profiles with a higher number of segments highlight a higher extent of the stripping reaction, as demonstrated by the higher value of the $CO_2$ vapor molar fraction along the column.

This result has important implication both at design and dynamic level and it was obtained thanks to the better discretization of the axial domain that led to a better evaluation of the internal fluxes in the stripper. In fact, remembering the definition of CSTR, a more detailed description permits to evaluate the internal fluxes in more points, and this generate more precise results. This fact is further corroborated by the analysis of the $CO_2$ and $H_2O$ interphase molar flow rate profiles reported in Fig. 4.4.

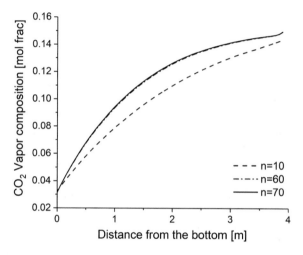

**Fig. 4.3** Stripper $CO_2$ vapor composition profile variation with the number of segments for Run 1

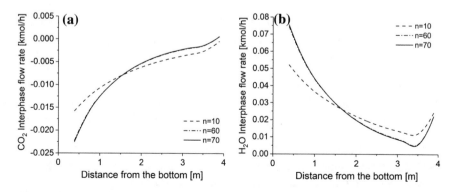

**Fig. 4.4** Stripper **a** $CO_2$ and **b** $H_2O$ interphase molar flow rate profiles variation with the number of segments for Run 1

It can clearly be seen that the profiles obtained with 10 segments are significantly different from those at 70 segments, showing the influence of the number of segments on the system solution and justifying this analysis. Moreover, the profile with 70 segments allows to identify an initial part of the column where the absorption process takes place over the stripping one. The process is characterized by an interphase transfer of water from the vapor to the liquid phase throughout the column, as a consequence of the condensation due to the endothermic stripping reactions.

This result has evident implication in design, because an inadequate discretization of the axial domain can lead to an under-/over-estimation of the column dimensions and required duty, and in the assessment of a control system structure, because the profiles would be different from the real ones.

## 4.6   UTA Plant: Run 47

After the validation of the model for what concerns the SINTEF plant, the stripping section of the University of Texas at Austin facility was considered. As already mentioned in Sect. 4.2, in the case of Run 47 the condenser temperature and the $CO_2$ output gas mass flow rate were fixed to saturate the degrees of freedom. This choice was made to highlight the effect of the number of segments analysis in the evaluation of the reboiler duty. It must be noticed that in this case, since the classic stripper configuration with reflux from the condenser was involved, both condenser and reboiler were modeled within the *RadFrac*$^{TM}$ model. Differently from the previous case studied, it was not possible to fix the chosen degrees of freedom directly. In particular, the mass reflux ration was used to fix the condenser temperature, while the total gas molar flow rate was used to fix the $CO_2$ out gas mass flow rate by means of the *Design Specification* tool included in the *RadFrac*$^{TM}$ block. The plant was simulated varying the number of segments and the results are reported in Fig. 4.5.

The flat part in the middle of the profile corresponds to the redistribution zone. Since the column is higher than the SINTEF plant one, the asymptotic behavior is reached only after 90 segments. The comparison between the experimental and the model profiles for what concerns the liquid temperature is reported in Fig. 4.6.

As it is possible to observe from Fig. 4.6, the model is able to describe adequately the experimental data, as furtherly demonstrated by the value of the standard error, which is about 0.7 K when 90 segments are used.

For what concerns the $CO_2$ vapor composition, the variation of the profiles with the number of segments is showed in Fig. 4.7.

Similarly to Run 1 from the SINTEF plant, the model with a higher number of segments always highlights a higher extent of the stripping reaction. Again, this is a consequence of the better discretization of the axial domain.

Table 4.11 summarizes the stripper performance with the variation of the number of segments compared with the experimental data.

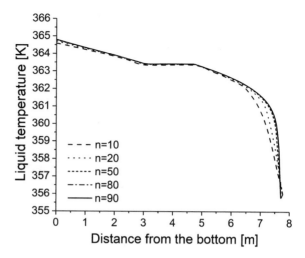

**Fig. 4.5** Stripper liquid temperature profile variation with the number of segments for Run 47

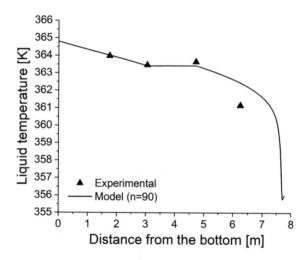

**Fig. 4.6** Comparison between the model liquid temperature profile and the experimental data for Run 47

From the analysis of Table 4.11 different conclusions can be made:

- the chosen set of degrees of freedom fixes the output conditions on the lean solvent and the gas from the condenser. For this reason, there is no difference in the output values with the variation of the number of segments. At the same time, the reboiler temperature is not influenced by the parameter variation;
- the number of segments has a significant effect on the value of the mass reflux ratio and the reboiler duty. Since the output conditions are fixed by the degrees of freedom, what changes is the internal behavior of the column. From the analysis of Fig. 4.7, it was found that the $CO_2$ molar fraction value with 10 segments is always the lowest one all along the column, with the highest difference at the top of the column. But since the output conditions are fixed, in

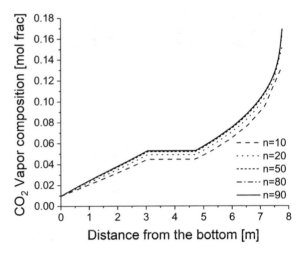

**Fig. 4.7** Stripper $CO_2$ vapor composition profile variation with the number of segments for Run 47

**Table 4.11** Stripper performance for Run 47

| Performance | Experimental | Number of segments | | | |
|---|---|---|---|---|---|
| | | 10 | 20 | 80 | 90 |
| Mass reflux ratio | | 2.57 | 2.21 | 1.95 | 1.94 |
| *Lean solvent* | | | | | |
| Loading (mol $CO_2$/mol MEA) | 0.286 | 0.268 | 0.268 | 0.268 | 0.268 |
| *Gas from condenser* | | | | | |
| Mixture flow rate (kg/h) | – | 94.89 | 94.89 | 94.89 | 94.89 |
| $CO_2$ flow rate (kg/h) | 92 | 91.8 | 91.8 | 91.8 | 91.8 |
| *Reboiler* | | | | | |
| Temperature (K) | – | 365.4 | 365.4 | 365.4 | 365.4 |
| Duty (kW) | 205 | 248.8 | 224.2 | 206.8 | 206.1 |

order to respect the constraint, the simulator finds that a high reboiler duty is needed to have the imposed value of the $CO_2$ mass flow rate from the condenser. For the same reason, the reflux ratio must be high. This is a consequence of the poor description of the column internal fluxes. On the other hand, when a proper number of segments is used, the internal fluxes are described correctly, the amount of $CO_2$ at the top of the column is higher and then the value of the reboiler duty and the mass reflux ratio are significantly reduced. In particular, for what concerns the reboiler duty, Table 4.12 shows how varying the number of segments it is possible to reduce considerably the error in the estimation of this fundamental variable.

This result has fundamental implications in the design of the stripper since, once the process targets are fixed, the estimation of the reboiler duty represents a crucial point to quantify the energy requirement. Furthermore, the vapor flow produced in the reboiler influences the column and the equipment dimensions.

**Table 4.12** Variation of the error in the evaluation of the reboiler duty with the number of segments

| Performance | Experimental | Number of segments | | | |
|---|---|---|---|---|---|
| | | 10 | 20 | 80 | 90 |
| Reboiler duty (kW) | 205 | 248.8 | 224.2 | 206.8 | 206.1 |
| Error (%) | – | 21.4 | 9.3 | 0.9 | 0.5 |

# References

1. Tobiesen FA, Juliussen O, Svendsen HF (2008) Experimental validation of a rigorous desorber model for $CO_2$ post-combustion capture. Chem Eng Sci 63(10):2641–2656
2. Dugas RE (2006) Pilot plant study of carbon dioxide capture by aqueous monoethanolamine. Dissertation, The University of Texas at Austin
3. Errico M, Madeddu C, Pinna D et al (2016) Model calibration for the carbon dioxide-amine absorption system. Appl Energy 183:958–968
4. Madeddu C, Errico M, Baratti R (2017) Rigorous modeling of a $CO_2$-MEA stripping system. Chem Eng Trans 57:451–456

# Chapter 5
# Absorption Section Design Analysis

In this chapter, the design of the absorption section of an industrial $CO_2$ post-combustion capture system using MEA as solvent is analyzed. After the process description, the gas and liquid feed streams are characterized. A two-steps procedure is adopted for the design. Initially, the minimum number of units and the minimum solvent flow rate are determined, then the role of the temperature bulge in the absorber design is discussed. The influence of the molar L/V ratio, which affects the amount of solvent to be used in the process, is studied by means of the analysis of the liquid temperature profiles. Then, the effective solvent flow rate and column dimensions are evaluated. The proposed design procedure for the absorber is proved to avoid the presence of isothermal zones in the column, guarantying the use of the entire packing.

## 5.1 Introduction to the Design of an Industrial $CO_2$-MEA Reactive Absorption Plant

The model developed in Chap. 2 was validated for the absorption and the stripping sections in Chaps. 3 and 4, respectively. The validated model is then used in this chapter with the aim of analyzing the design of an industrial-scale plant. In general, the works on the modeling of the $CO_2$ capture by means of reactive absorption-stripping with MEA have been concentrated on the model validation using experimental data from pilot-plant facilities [1–7]. For what concerns the industrial plants, very few experimental data sets are available, mostly reporting values at the extremes of the columns only [8]. This fact makes particularly difficult to test models at this scale.

© The Author(s), under exclusive license to Springer Nature Switzerland AG 2019    55
C. Madeddu et al., *CO2 Capture by Reactive Absorption-Stripping*,
SpringerBriefs in Energy, https://doi.org/10.1007/978-3-030-04579-1_5

Nevertheless, different works have been focused on industrial plants. For example, an economic comparison between a post-combustion capture plant with MEA and an $O_2/CO_2$ recycle combustion plant was made by Singh et al. [9]. Alie et al. [10] and Abu-Zahra [11] performed a series of sensitivity analysis to study the effect of different operating parameters. Cau et al. [12] investigated the effect of the $CO_2$ capture process on an Ultra Super Critical (USC) steam system and an Integrated Gasification Combined Cycle (IGCC) plant. A dynamic model was used by Lawal et al. [13] and Nittaya et al. [14] for the design of a $CO_2$ post-combustion capture plant using MEA as solvent.

In most cases, there is no agreement between the results obtained by the different researchers. Several reasons can explain this fact:

- Differences in the amount of gas treated and in the operating conditions, which influence the determination of the optimal operating parameters. For example, a range between 0.25 [10] and 0.32 [11] mol $CO_2$/mol MEA was found for the optimal lean solvent loading.
- Focus on the effect of few operating parameters only: Lawal et al. [13] and Nittaya et al. [14] investigated the effect of the absorber packing height on the energy consumption, while Cau et al. [12] reported the variation of the $CO_2$ removal with the L/V ratio and the reboiler duty. Only Abu-Zahra et al. [11] made a sensitivity analysis varying all the most important operating parameters;
- Use of different mathematical models, from the equilibrium stages model [10–12] to the rate-based model [13, 14]. The choice of a simple model over a rigorous one can lead to significant differences in the results.

The typical approach for the design of the reactive absorption-stripping processes involves a series of sensitivity analyses, by which the column dimensions and the operating parameters are varied over a certain range of values in order to obtain the desired final performance. In this kind of analyses the column profiles are usually neglected. This fact could lead to columns that do not operate correctly, although they respect the results at the extremes, i.e., product purity, removal percentage, etc.

A process analysis and design procedure based on the contemporary focus on the internal column profiles and the performance values is reported in the present chapter. After the characterization of the gas and the liquid feed streams, the effect of various important parameters is investigated. In particular, for the absorber the effect of the L/V ratio is considered, determining the limits where the column is consistent both from the output streams and the internal behavior standpoints.

The main aim of this analysis is to show how the choice of the operating parameters influences the design and the operating conditions of the absorber and the stripper. In particular, it is proved that a design based only on the global performance values may lead to columns that do not operate correctly. This aspect is highlighted by the simultaneous analysis of the operating parameters and their impact on the column temperature and composition profiles [15].

## 5.2   Process Description

The absorption-solvent regeneration process reported in Fig. 5.1 is considered for the process analysis. The plant is divided in two interconnected sections, i.e., the absorption, where the $CO_2$ is transferred from the gaseous to the liquid phase, and the stripping, where the mass separation agent is recovered. In the absorption section, the $CO_2$-rich flue gas is sent to the bottom of the absorber, where it flows in a countercurrent arrangement with the lean liquid solvent. The exhaust gas exits the top of the column and it is sent to the stack.

The $CO_2$-rich solvent from the bottom of the absorber is pumped to a cross heat-exchanger, where its temperature is increased, and then to the top of the stripper. The liquid flows countercurrent with the vapor flow generated by the reboiler. In this case, the $CO_2$ is transferred from the liquid to the vapor phase. From the top of the stripper, a gaseous stream composed mainly by $CO_2$ and $H_2O$ enters a partial condenser where the $CO_2$ is concentrated in the gas phase. The gas is then sent to compression and subsequent storage or re-utilization. The lean solvent exiting the stripper bottom is partly vaporized in the reboiler and then sent to the heat-exchanger where it supplies its sensible heat to the stripper feed. The solvent is mixed with the water recovered from the condenser, and then furtherly cooled and recycled to the top of the absorber. This configuration differs from the classic absorption-solvent regeneration process [16, 17], since the water recovered is not recycled back to the top of the stripper as reflux. As it is going to be explained in detail in Chap. 6, this choice is motivated by energy saving reasons.

Both the absorber and the stripper are packed columns, chosen over the plate ones because the packing provides a higher contact area and less pressure drop. The chosen packing for both the column is Sulzer Mellapak 252Y. Though mono-ethanolamine is used as solvent, as it is the most studied and proven to be the most

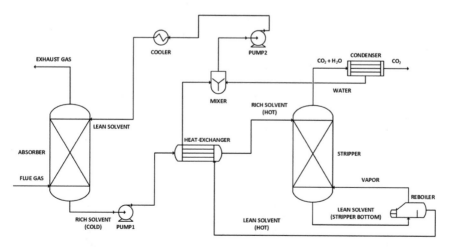

**Fig. 5.1**  Industrial $CO_2$ post-combustion capture by reactive absorption-stripping plant flowsheet

mature one for this process [3, 18–20], this design analysis is independent of the solvent used. The two columns are considered separately following the flowsheet decomposition approach proposed in the work of Alie et al. [10]. This method consists in considering each column as standalone, and then using the results as initial guesses for the coupled system analysis. Typically, although no general rule is present in the literature, packed absorption and stripping column present packing height/diameter ratio values larger than 1 [14]. For this reason, it was decided to impose that both absorber and stripper must have a packing height/diameter ratio higher or equal to 1.

## 5.3  Feed Streams Characterization

The purpose of the absorption section is to remove the $CO_2$ from the flue gas entering the capture plant. It is assumed that other pollutants ($SO_2$, $NO_x$, etc.) have been previously removed. Typically, before being sent to the stack, the exhaust gas from the top of the absorber goes to a waterwash section, where part of the vaporized solvent is recovered. This section is not considered in this Chapter since it does not influence the scope of the analysis. The waterwash section will be considered in Chap. 7 for the economic evaluation of the plant.

The flue gas composition, temperature and pressure reported in the work of Lawal et al. [13] for a coal-fired sub-critical power plant are considered as reference. Differently from Lawal et al., where the flue gas flow rate was representative of a 500 MWe power plant, in this case it was decided to deal with a 250 MWe power plant, as this is the reference target defined by the European Community for demo-scale $CO_2$ capture plants [21]. The complete flue gas characterization is reported in Table 5.1.

A $CO_2$ removal efficiency of 90% is fixed, in agreement with different works [9, 11–13, 21, 22]. In order to achieve this target, a 30 wt% MEA aqueous solution is used [11–13, 21, 22]. Furthermore, the temperature of the solvent is defined at 308.15 K, assuming the presence of cooling water at 298.15 K entering the cooler placed before the absorber. The lean solvent loading value is fixed at 0.3, which is an average value among those that can be found in the literature, as reported in Table 5.2. A more extended analysis with the variation of the lean solvent loading is reported in the work of Madeddu et al. [15].

**Table 5.1**  Flue gas characterization

| Variable | Value |
| --- | --- |
| Mass flow rate (kg/s) | 300 |
| Temperature (K) | 313.15 |
| Pressure (bar) | 1 |
| $CO_2$ (mass frac) | 0.21 |
| $H_2O$ (mass frac) | 0.042 |
| $N_2$ (mass frac) | 0.748 |

**Table 5.2** Brief literature review on the lean solvent loading values

| Reference | Loading (mol $CO_2$/mol MEA) |
| --- | --- |
| [11] | 0.32 |
| [12] | 0.28 |
| [13] | 0.29 |
| [14] | 0.3 |
| [21] | 0.271 |

**Table 5.3** Lean solvent characterization

| Variable | Value |
| --- | --- |
| Temperature (K) | 313.15 |
| Pressure (bar) | 1 |
| $CO_2$ (mass frac) | 0.0649 |
| $H_2O$ (mass frac) | 0.6351 |
| MEA (mass frac) | 0.3 |

The characterization of the lean solvent is reported in Table 5.3. It must be noted that the flow rate is not reported since it is evaluated in the next sections.

## 5.4   Absorber Analysis and Design Implications

The objective of the absorber design is the determination of the amount of solvent and the column dimensions, i.e., the packing height and the diameter. The procedure adopted consists of two steps:

1. Evaluation of the number of absorption units required and the minimum solvent flow rate with an infinite packing height;
2. Evaluation of the effective packing height with different solvent flow rate values.

The two steps are described in detail in Sects. 5.4.1–5.4.3.

### 5.4.1   Evaluation of the Minimum Number of Absorbers and the Minimum Solvent Flow Rate

In this section, the minimum solvent flow rate is determined using an *infinite* packing height. Then, the packing height was fixed at 100 m, using an approach similar to that used for the plate columns, where a high (theoretically infinite) number of stages is set to determine the minimum reflux ratio [16, 17].

In the capture plant, the absorption section is characterized by the highest column diameters, due to the large amount of gas involved in the process. Different works in the literature reported that for this kind of absorbers the diameter should not be higher than 12 m [9, 13, 14]. For this reason, the minimum number of units is determined in this first step, since it is intrinsically related to the column diameter.

When the design of a packed column is considered using the *RadFrac*$^{TM}$ model, the *Packing Rating—Design Mode* option must be activated to determine the column diameter. In this case two parameters must be specified:

- **Base flood**. It corresponds to the maximum percentage of flooding velocity allowed for the evaluation of the column diameter. In this work, a gas velocity of 70% of the flooding velocity is fixed;
- **Base stage**. The evaluation of the column diameter is performed with reference to a specific point in the column. Usually, this point corresponds to the part of the column which is more stressed, i.e., the point where the gas/vapor flow rate reaches its maximum value. As the absorption process is characterized by exothermic reactions that cause water vaporization, the maximum vapor flow rate will most likely not be neither at the top or the bottom of the absorber, but at some point within the column, where most of the reactions happen. To determine this specific point in the column, the *Packing Sizing* tool, included in the *RadFrac*$^{TM}$ model, is used. It must be highlighted that in the case of the rate-based model the base stage should be more appropriately referred to as *Base Segment*.

For what concerns the simulations, a model with 160 segments is used for the *infinite* packing height column. This value was chosen based on the norm of the difference between the interphase $CO_2$ molar flow vector evaluated at 150 and 160 segments, which was in the order of $10^{-3}$. A further increase in the number of segments would lead to an excessive computational cost, which is unnecessary for the scope of the analysis. The same criterion is used for all the subsequent simulations to determine the number of segments for the system solution.

With respect to the minimum number of absorption units, it is found that one absorber leads to diameters close to or higher than the imposed limit of 12 m with the infinite packing height. In particular, a diameter of 13.1 m was found for the case studied. Moreover, considering the effective column, the diameter is certainly going to increase due to the higher solvent flow rate involved. For this reason, it can be concluded that at least two units are needed to respect the constraint on the absorber diameter. Consequently, the flue gas stream must be divided into two equal parts that flow into two identical absorbers. At this point, to achieve the target of 90% removal of $CO_2$, 28.35 kg/s of $CO_2$ must be removed from the flue gas in each absorber.

After the determination of the minimum number of absorption units, the minimum solvent flow rate is evaluated using the *Model Analysis Tools—Sensitivity* present in Aspen Plus$^{®}$. In particular, the solvent flow rate is varied until the desired removal of $CO_2$ is achieved. The results are reported in Table 5.4, together with the rich solvent characterization and the column features.

**Table 5.4** Results for the infinite packing height column analysis

| Variable | Value |
|---|---|
| *Lean solvent* | |
| Flow rate (kg/s) | 487.2 |
| *Rich solvent* | |
| Flow rate (kg/s) | 503.7 |
| Temperature (K) | 317 |
| $CO_2$ (mol frac) | 0.0672 |
| MEA (mol frac) | 0.1179 |
| $H_2O$ (mol frac) | 0.8149 |
| Loading (mol $CO_2$/mol MEA) | 0.569 |
| *Column features* | |
| Height (m) | 100 |
| Diameter (m) | 9.28 |
| Base stage (segment) (m) | 6 |
| L/V ratio (kmol/kmol) | 3.86 |

## 5.4.2 The Role of the Temperature Bulge in the Absorber Design

Once the minimum solvent flow rate is determined, it is possible to shift from the infinite packing height column to the effective packing height one. The first step of this procedure is the evaluation of the effective solvent flow rate which is, in general, a multiple of the minimum solvent flow rate. As reported by Seader et al. [16], the effective solvent flow rate can be computed as (Eq. 5.1):

$$L_0^{eff} = (1 \div 2)L_0^{min} \tag{5.1}$$

In the case of a $CO_2$-MEA absorber this computation is not immediate. This is due to the fact that in this kind of system, as already discussed in Chap. 3, the molar L/V ratio influences the position of the typical bulge in the temperature profiles. In particular, three situations are possible:

1. **L/V < L/V$_{low}$:** the bulge is positioned at the top of the absorber; the minimum driving force is present at the bottom of the column. The bulge does not affect the performance of the process.
2. **L/V$_{low}$ < L/V < L/V$_{up}$:** the temperature profiles do not present a clear bulge neither at the top or the bottom of the absorber; the shape of the curve shows a soft bulge distributed all along the column. The minimum driving force tends to appear somewhere in the middle of the column. In this case, the temperature bulge influences the performance of the absorber. Furthermore, in this situation the temperature gradient is very small along the column and close to be zero in the middle. This is an indication that the absorber is not working correctly, as a large part of the column is practically isothermal, and then not correctly used.

3. **L/V > L/V$_{up}$**: the bulge is placed at the bottom of the column; the minimum driving force is present at the top of the absorber. The temperature bulge does not affect the performance of the absorber.

The values of the extremes of the L/V ratio interval vary based on the examined case. In the plant investigated by Kvasmdal and Rochelle [23], the values of L/V$_{low}$ and L/V$_{up}$ were found to be around 5 and 6, respectively.

In the light of what was said above, in the case of the absorber design, it is necessary to avoid value of the effective solvent flow rate that lead to an L/V ratio value inside category 2. The temperature profiles, being closely related to the L/V ratio, highlight the portion of the column where the temperature gradient is close to zero, giving important indications on the quality of the process and should always be checked together with the outputs.

## 5.4.3   Evaluation of the Effective Solvent Flow Rate and the Effective Column Dimensions

### 5.4.3.1   L/V Ratio Analysis

The value of the L/V ratio must be examined for different multiples of the minimum solvent flow rate. According to the values reported by Kvamsdal and Rochelle [23], it can be found that the molar L/V ratio is inside Category 2 for a value of $L_0^{eff}$ in the range $(1.3 \div 1.5)L_0^{min}$. Then, for multiples of the minimum solvent flow rate that lead to an L/V ratio in this interval it is expected to have a mild temperature bulge in the absorber. However, since the values reported by Kvamsdal and Rochelle were evaluated in a different pilot-plant facility, this interval is purely indicative, and a deeper investigation based on the analysis of the column profiles must be conducted.

### 5.4.3.2   Absorber Liquid Temperature Profiles

The value of the L/V ratio influences the temperature profiles and limits the choice of the effective solvent flow rate. This effect can be studied analyzing the liquid temperature profiles, which are reported in Fig. 5.2a for different values of the effective solvent flow rate to corroborate the correctness of the L/V ratio analysis. At the same time, Fig. 5.2b shows the corresponding $CO_2$ vapor composition profiles. Since for each multiple of the minimum solvent flow rate the effective packing height is different, the relative distance from the bottom of the column is used for the x-axis. For all the subsequent simulations of the absorber, 100 discretization segments were found to be sufficient to have a correct mathematical solution.

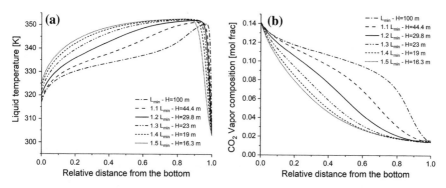

**Fig. 5.2** Variation of the **a** liquid temperature and **b** $CO_2$ vapor composition profiles for different multiples of the minimum solvent flow rate

As it can be observed from Fig. 5.2a, the temperature profile tends to become flatter with the increase of $L_0^{eff}$, i.e., with the increase of the L/V ratio. In other words, the temperature bulge becomes less evident with the increase of the solvent flow rate. It can be noticed that moving from $L_0^{eff} = 1.2\ L_0^{min}$ to $L_0^{eff} = 1.3\ L_0^{min}$ there is the transition from Category 1 to Category 2. This is highlighted by the fact that an isothermal zone appears between a relative distance from the absorber bottom of 0.7 and 0.9 when $L_0^{eff} = 1.3\ L_0^{min}$. For higher multiples of the minimum solvent flow rate this situation is even more evident, since a larger isothermal zone appears. This result is furtherly corroborated by the analysis of the corresponding $CO_2$ vapor composition profiles, reported in Fig. 5.2b, where it can be observed that the increase of the effective solvent flow rate leads to have a portion of the column where the concentration gradient is close to zero. This fact indicates that the absorber is not working correctly, as in the isothermal part of the column the $CO_2$ interphase transfer is close to zero.

### 5.4.3.3  Absorber Dimensions

For each multiple of the minimum solvent flow rate reported in Fig. 5.2 the absorber dimensions, i.e., packing height and diameter, were evaluated as reported in Table 5.5. It must be noticed that it was possible to investigate only up to a multiple of the minimum solvent flow rate of 1.5. This was because for higher values of the effective solvent flow rate, using two absorption units, the column would have packing height/diameter ratio values smaller than 1. In order to study the behavior of the system for higher multiples of the minimum solvent flow rate, a higher number of units would be needed, since in that case there would be a lower of volume of gas to be treated per absorption units and, consequently, smaller diameters.

| Variable | Multiple of $L_0^{min}$ | | | | |
|---|---|---|---|---|---|
| | 1.1 | 1.2 | 1.3 | 1.4 | 1.5 |
| Packing height (m) | 44.4 | 29.8 | 23 | 19 | 16.3 |
| Diameter (m) | 9.47 | 9.67 | 9.83 | 9.97 | 10.1 |
| Base segment | 7 | 9 | 11 | 15 | 16 |

**Table 5.5** Absorption column features for different values of the effective solvent flow rate

From the analysis of Table 5.5, it can be noticed that the increase of the effective solvent flow rate leads to smaller packing heights and higher diameters.

According to the analysis of the liquid temperature profiles, only the cases for $L_0^{eff} = 1.1 \, L_0^{min}$ and $L_0^{eff} = 1.2 \, L_0^{min}$ ensure the absence of isothermal zones in the column. In order to choose the best case among the two, an optimization problem should be solved to find the optimal effective packing height and effective solvent flow rate, but this is not the scope of this analysis and then it is not considered in this work. In the next Chapter regarding the analysis of the stripper design the rich solvent obtained with the case at $L_0^{eff} = 1.2 \, L_0^{min}$ is used as feed in the stripping column. It must be pointed out that this case is chosen as example in this book.

## 5.5  Absorber Design Procedure Summary

In this section, the steps involved in the design of the absorber according to the procedure proposed are resumed:

I. Characterization of the flue gas and the liquid solvent (except for the flow rate) and definition of the target performance ($CO_2$ removal)
II. Infinite packing height column

1. Definition of the minimum number of absorption units to respect the constraint on the maximum value of the diameter;
2. Evaluation of the minimum solvent flow rate by means of a sensitivity analysis;

III. Effective column

1. Simulation of the absorber for different multiples of the minimum solvent flow rate and evaluation of the corresponding column dimensions (packing height and diameter);
2. Identification of the extremes of the L/V ratio where isothermal packing zones are avoided;
3. Choice of the effective solvent flow rate and effective packing height by means of an optimization problem.

# References

1. Tobiesen FA, Svendsen HF (2007) Experimental validation of a rigorous absorber model for $CO_2$ postcombustion capture. AIChE J 53(4):846–865
2. Lawal A, Wang M, Stephenson P et al (2009) Dynamic modelling of $CO_2$ absorption for post-combustion capture in coal-fired power plant. Fuel 88(12):2455–2462
3. Plaza JM, Wagener DV, Rochelle GT (2009) Modeling $CO_2$ capture with aqueous monoethanolamine. Energy Procedia 1(1):1171–1178
4. Zhang Y, Chen H, Chen C-C et al (2009) Rate-based process modeling study of $CO_2$ capture with aqueous monoethanolamine solution. Ind Eng Chem Res 48(20):9233–9246
5. Tobiesen FA, Juliussen O, Svendsen HF (2008) Experimental validation of a rigorous desorber model for $CO_2$ post-combustion capture. Chem Eng Sci 63(10):2641–2656
6. Tontiwachwuthikul P, Meisen A, Lim CJ (1992) $CO_2$ absorption by NaOH, monoethanolamine and 2-amino-2-methyl-1-propanol solutions in a packed column. Chem Eng Sci 47(2):381–390
7. Mac Dowell N, Samsatli NJ, Shah N (2013) Dynamic modelling and analysis of an amine-based post-combustion $CO_2$ capture absorption column. Int J Greenhouse Gas Control 12:247–258
8. Pintola T, Tontiwachwuthikul P, Meisen A (1993) Simulation of pilot-plant and industrial $CO_2$-MEA absorbers. Gas Sep Purif 7(1):47–52
9. Singh D, Croiset E, Douglas PL et al (2003) Techno-economic study of $CO_2$ capture from an existing coal-fired power plant: MEA scrubbing vs $O_2/CO_2$ recycle combustion. Energy Convers Manage 44(19):3073–3091
10. Alie C, Backham L, Croiset E et al (2005) Simulation of $CO_2$ capture using MEA scrubbing: a flowsheet decomposition method. Energy Convers Manage 46(3):475–487
11. Abu-Zahra MRM, Schneiders LHJ, Niederer JPM et al (2007) $CO_2$ capture from power plants: Part I. A parametric study of the technical performance based on monoethanolamine. Int J Greenhouse Gas. Control 1(1):37–46
12. Cau G, Tola V, Deiana P (2014) Comparative performance assessment of USC and IGCC power plants integrated with $CO_2$ capture systems. Fuel 116:820–833
13. Lawal A, Wang M, Stephenson P et al (2012) Demonstrating full-scale post-combustions $CO_2$ for coal-fired power plants through dynamic modelling and simulation. Fuel 101:115–128
14. Nittaya T, Douglas PL, Croiset E et al (2013) Dynamic modeling and evaluation of an industrial-scale $CO_2$ capture plant using monoethanolamine absorption processes. Ind Eng Chem Res 53(28):11411–11426
15. Madeddu C, Errico M, Baratti R (2018) Process analysis for the carbon dioxide chemical absorption-regeneration system. Appl Energy 215:532–542
16. Seader JD, Henley EJ, Koper DK (2010) Separation process principles: chemical and biochemical operations. Wiley, New York
17. Sinnott RK (2005) Coulson & Richardson's chemical engineering volume 6—chemical engineering design. Elsevier Butterworth-Heinemann
18. Wang M, Lawal P, Stephenson P et al (2011) Post-combustion $CO_2$ capture with chemical absorption: a state-of-the-art review. Chem Eng Res Des 89(9):1609–1624
19. Tan LS, Shariff M, Lau KK et al (2012) Factors affecting $CO_2$ absorption efficiency in packed column: a review. J Ind Eng Chem 18(6):1874–1883
20. Bui M, Gunawan I, Verheyen V et al (2014) Dynamic modelling and optimization of flexible operation in post-combustion $CO_2$ capture plants—a review. Comput Chem Eng 61:245–265
21. de Miguel Mercader F, Magneschi G, Fernander ES et al (2012) Integration between a demo size post-combustion $CO_2$ capture and full size plant. An integral approach on energy penalty for different process options. Int J Greenhouse Gas Control 11S:S102–S113
22. Lin Y-J, Wong DS-H, Jang S-S (2012) Control strategies for flexible operation of power plant with $CO_2$ capture plant. AIChE J 58(9):2697–2704
23. Kvamsdal HM, Rochelle GT (2008) Effect of the temperature bulge in $CO_2$ absorption from flue gas by aqueous monoethanolamine. Ind Eng Chem Res 47(3):867–875

# Chapter 6
# Stripping Section Design Analysis

The analysis of the stripping section design procedure of an industrial $CO_2$-MEA post-combustion capture system is considered in this chapter. An alternative plant configuration without reflux is adopted with the aim of reducing the consumption of steam in the reboiler. Then, the most important operating parameters are described in detail. After the rich solvent characterization, the effect of the packing height on the reboiler duty and the column diameter is analyzed. Due to the impossibility to use the classic minimum and effective stripping agent design approach, a criterion for the definition of the stripper packing height based on the analysis of the liquid temperature gradient profiles is proposed.

## 6.1 Introduction to the Design of an Industrial $CO_2$-MEA Reactive Stripping Plant

The stripping section represents, from an economic point of view, the most crucial one, due to the energy consumed in the reboiler. In fact, part of the steam generated by the power plant is used as auxiliary fluid in the reboiler decreasing the thermal efficiency of the entire capture plant [1, 2]. Up to now, different works were focused on finding the operating conditions that minimize the reboiler duty or developing new plant schemes to improve the energy utilization by means of Process Integration techniques [3]. Similarly to the absorber design case, also for the stripper conflicting results are reported in the literature. For instance, Singh et al. found an optimum specific reboiler duty of 1.7 GJ/tCO$_2$ [4], while Nittaya et al. found the same optimum value to be 4.1 GJ/tCO$_2$ [5]. These discrepancies are mainly due to differences in the modeling approach or in the operating condition as discussed in Chap. 5 for the absorber. In this Chapter, a design approach based on the contemporary focus on the internal column profiles and the final performance is adopted for the stripper. In particular, an alternative column configuration without

© The Author(s), under exclusive license to Springer Nature Switzerland AG 2019
C. Madeddu et al., *CO$_2$ Capture by Reactive Absorption-Stripping*,
SpringerBriefs in Energy, https://doi.org/10.1007/978-3-030-04579-1_6

reflux from the condenser is used to avoid an unnecessary energy consumption. A criterion for the determination of the minimum packing height is proposed through the analysis of the liquid temperature and temperature gradient profiles [6].

## 6.2  Stripper Configuration

The role of the stripping section is the regeneration of the $CO_2$-rich solvent which arrives from the absorber. In this case the separating agent is the vapor flow rate produced in the reboiler, since the reactions in the stripping process are endothermic. Furthermore, in the case of the $CO_2$-MEA system, as carbon dioxide is a gas, its mass transfer from the liquid to the vapor phase is no-heat consuming. Then, the duty is needed only to reverse the absorption reactions. In general, the process must be operated at the highest possible temperature once the feed conditions are fixed. In order to achieve this objective, the configuration of the stripper was examined before the design analysis. Different works dealing with the stripper design problem reported the possibility to send part of the water recovered in the condenser back to column top as reflux [4, 5, 7–12]. Since the reflux would enter the column at a considerably lower temperature compared to the stripper feed, this configuration would lead to a decrease in the column top temperature. Consequently, there would be the need for a higher duty from the reboiler to heat the cold reflux and an increase in the column diameter. For this reason, as reported in Sect. 5.2, it is proposed to mix the cold water recovered in the condenser with the lean solvent exiting from the heat-exchanger (Fig. 6.1), as it was previously done in the work by Oexmann and Kather [13].

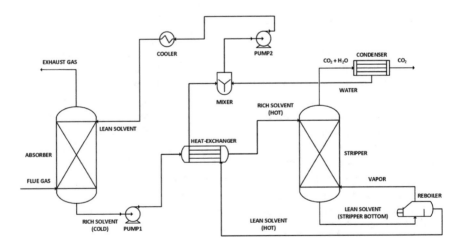

**Fig. 6.1** Industrial $CO_2$ post-combustion capture by reactive absorption-stripping plant flowsheet

## 6.3  Stripper Operating Conditions

After the choice of the stripper configuration, it is necessary to define several operating parameters that affect the performance of the process before the design. These parameters are discussed separately in the next sections.

### 6.3.1  Stripper Pressure

Typically, the stripper operates at a pressure higher than the atmospheric one. This is due to the fact that, according to the work of Freguia [14], the heat of absorption of $CO_2$ in MEA is almost two times the water heat of vaporization. Then, according to the Clausius-Clapeyron equation, the $CO_2$ vapor pressure increase more rapidly compared to the $H_2O$ vapor pressure. For this reason, the stripper must work at higher pressure compared to the absorber, in order to reach higher temperatures and favor the transfer of $CO_2$ over water. However, there is a limit on the pressure value which is imposed by the degradation temperature of MEA. This value was set at 495.15 K according to Alie [15], though Davis and Rochelle highlighted that the degradation is minor when the reboiler temperature is held below 383.15 K [16]. For this reason, the column pressure must be set at the highest value that guarantees a solvent boiling temperature lower than the solvent degradation temperature. For a 30 wt% MEA aqueous solution this pressure corresponds to 1.8 bar, and this value was fixed in the stripper.

### 6.3.2  Condenser Temperature

A gaseous mixture containing mainly $CO_2$ and $H_2O$ exits the top of the stripper. This stream is sent to the partial condenser, where the $CO_2$ is concentrated in the gas phase while the water is recovered in the liquid phase. Assuming the availability of cooling water at 298.15 K , the condenser temperature is set to 308.15 K . With this specification, a concentration of $CO_2$ in the gas of 96 mol% or higher is ensured before the compressor.

### 6.3.3  Stripper Performance

The stripper should be designed to remove the amount of $CO_2$ captured in the absorber. Removing a higher quantity would lead to a higher energy consumption. On the other hand, removing a lower amount would lead to a dirtier solvent and, consequently, a higher lean solvent flow rate in the absorber to ensure the 90% $CO_2$ removal.

## 6.4   Stripper Analysis and Design Implications

### 6.4.1   Rich Solvent Characterization

As already mentioned at the end of Sect. 5.4.3.3, the case corresponding to $L_0^{eff} = 1.2\ L_0^{min}$ is considered for the design of the stripper. Table 6.1 reports the characterization of the rich solvent obtained before the pump.

Before entering the stripper, the pressure of the rich solvent is increased to 1.8 bar, while the temperature is initially fixed to the boiling point, since it is typical for the stripper to send the feed as a saturated liquid [9, 10].

### 6.4.2   Effect of the Packing Height

When the absorber was considered in Chap. 5, the approach for determining the column dimensions involved firstly the evaluation of the minimum solvent flow rate with an infinite packing height. Then, the effective column dimensions were computed using the effective solvent flow rate. In the case of the stripper, the corresponding procedure should involve the use of the minimum and the effective stripping agent. In the case of a $CO_2$-MEA stripping system this approach cannot be used. This is demonstrated by the results showed in Table 6.2, where the variation of the reboiler duty and the column diameter for different values of the packing height are reported.

From Table 6.2 it is possible to observe that the packing height has a low influence on both the reboiler duty and the column diameter values. This happens because, once the rich solvent is defined, the duty is used only to heat the feed and to reverse the absorption reactions, independently of the amount of contact surface available. At this point, an alternative procedure to determine the stripper packing height is needed. For this reason, the liquid temperature profiles were checked to investigate the column internal behavior varying the packing height. Similarly to the case of the absorber, due to the different packing heights considered, the relative distance from the bottom is reported in x-axis. For the simulations, a model with 70 segments is used. The liquid temperature profiles for different values of the packing height are reported in Fig. 6.2.

**Table 6.1**  Rich solvent characterization

| Variable | Value |
|---|---|
| Mass flow (kg/s) | 604.1 |
| Temperature (K) | 320 |
| Pressure (bar) | 1 |
| $CO_2$ (mol frac) | 0.0614 |
| MEA (mol frac) | 0.1171 |
| $H_2O$ (mol frac) | 0.8214 |

**Table 6.2** Variation of the reboiler duty and the column diameter for different values of the stripper packing height

| Packing height (m) | Reboiler duty (MW) | Diameter (m) |
|---|---|---|
| 3 | 145.7 | 6.93 |
| 4 | 145.7 | 6.92 |
| 5 | 145.7 | 6.91 |
| 6 | 145.8 | 6.91 |
| 7 | 145.8 | 6.9 |
| 8 | 145.8 | 6.89 |
| 9 | 145.8 | 6.89 |
| 10 | 145.8 | 6.88 |
| 11 | 145.8 | 6.88 |
| 12 | 145.8 | 6.87 |
| 13 | 145.8 | 6.87 |
| 14 | 145.8 | 6.86 |

From the analysis of Fig. 6.2 it can be noticed that the increase of the packing height leads to have a flat zone in the middle of the column. For this reason, the liquid temperature gradients, reported in Fig. 6.3 for the corresponding values of the packing height, can be analyzed in order to quantify this behavior.

In Fig. 6.3 it is evidenced that after a certain value of the packing height an extended zone where the temperature gradient is less than 1 K/m appears. This means that in that portion of packing the stripper can be considered isothermal. It is then decided to choose the maximum packing height at which the temperature gradient is always higher than 1 K/m [6]. This criterion, in analogy with the absorber case analyzed in Chap. 5, ensures the obtainment of a column where the whole packing height is used, avoiding isothermal zones that indicate the absence of reaction. In this specific case, a value of the packing height of 6 m is found. However, it must be highlighted that this value would violate the imposed constraint on the packing height/diameter ratio, which must be at least 1 or higher, as reported in Chap. 5. For this reason, 7 m of packing height are considered.

**Fig. 6.2** Variation of the liquid temperature profile for different packing height values

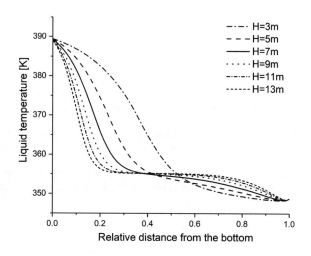

**Fig. 6.3** Variation of the
liquid temperature gradient
profile for different packing
height values

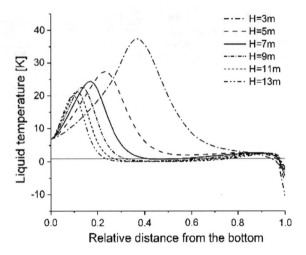

## 6.5  Stripper Design Procedure Summary

In this section, the steps involved in the design of the stripper according to the procedure proposed are resumed:

I. Characterization of the rich solvent and definition of the target performance in terms of amount of $CO_2$ stripped.
II. Effective column.

1. Simulation of the plant for different values of the stripper packing height;
2. Evaluation of the liquid temperature gradient profiles for the corresponding values of the packing height;
3. Choice of the packing height that guarantees to have a liquid temperature gradient higher than 1 K/m in each point of the column;
4. Computation of the reboiler duty and the column diameter.

## References

1. Rao AB, Rubin ES (2002) A technical, economic, and environmental assessment of amine-based $CO_2$ capture technology for power plant greenhouse gas control. Environ Sci Technol 36(20):4467–4475
2. Zhang Q, Turton R, Bhattacharyya D (2016) Development of model and model-predictive control of an MEA-based postcombustion $CO_2$ capture process. Ind Eng Chem Res 55 (5):1292–1308
3. Neveaux T, Le Moullec Y, Corriou JP et al (2013) Energy performance of $CO_2$ capture processes: interaction between process design and solvent. Chem Eng Trans 35:337–342

4. Singh D, Croiset E, Douglas PL et al (2003) Techno-economic study of $CO_2$ capture from an existing coal-fired power plant: MEA scrubbing vs $O_2/CO_2$ recycle combustion. Energy Convers Manage 44(19):3073–3091
5. Nittaya T, Douglas PL, Croiset E et al (2013) Dynamic modeling and evaluation of an industrial-scale $CO_2$ capture plant using monoethanolamine absorption processes. Ind Eng Chem Res 53(28):11411–11426
6. Madeddu C, Errico M, Baratti R (2018) Process analysis for the carbon dioxide chemical absorption-regeneration system. Appl Energy 215:532–542
7. Alie C, Backham L, Croiset E et al (2005) Simulation of $CO_2$ capture using MEA scrubbing: a flowsheet decomposition method. Energy Convers Manage 46(3):475–487
8. Abu-Zahra MRM, Schneiders LHJ, Niederer JPM et al (2007) $CO_2$ capture from power plants: Part I. A parametric study of the technical performance based on monoethanolamine. Int J Greenhouse Gas Control 1(1):37–46
9. Seader JD, Henley EJ, Koper DK (2010) Separation process principles: chemical and biochemical operations. Wiley, New York
10. Sinnott RK (2005) Coulson & Richardson's chemical engineering volume 6—Chemical engineering design. Elsevier Butterworth-Heinemann
11. de Miguel Mercader F, Magneschi G, Fernander ES et al (2012) Integration between a demo size post-combustion $CO_2$ capture and full size plant. An integral approach on energy penalty for different process options. Int J Greenhouse Gas Control 11S:S102–S113
12. Kang CA, Brandt AR, Durlofsky LJ et al (2016) Assessment of advanced solvent-based post-combustion $CO_2$ capture processes using a bi-objective optimization technique. Appl Energy 179:1209–1219
13. Oexmann J, Kather A (2009) Post-combustion $CO_2$ capture in coal-fired power plants: comparison of integrated chemical absorption processes with piperazine promoted potassium carbonate and MEA. Energy Procedia 1(1):799–806
14. Freguia S (2002) Modeling of $CO_2$ removal from flue gases with monoethanolamine. Dissertation, The University of Texas at Austin
15. Alie C (2004) $CO_2$ capture with MEA: integrating the absorption process and steam cycle of an existing coal-fired power plant. Dissertation, University of Waterloo
16. Davis J, Rochelle GT (2009) Thermal degradation of monoethanolamine at stripper conditions. Energy Procedia 1(1):327–333

# Chapter 7
# Complete Flowsheet and Economic Evaluation

In this chapter, the design of the cross heat-exchanger, which interconnects the absorption and the stripping section, is examined in the first place. In this context, the stripper feed temperature is discussed in relation to its influence on the reboiler duty. Then, the flowsheet is completed with the design of the section dedicated to the recovery of the solvent lost in the absorber exhaust gas and the introduction of the auxiliary equipment. The last part of the chapter is dedicated to the overall economic evaluation of the plant. In particular, the capital and the operating costs are determined and their impact on the total annual costs is discussed.

## 7.1 Introduction

Chapters 5 and 6 were dedicated to the design of the absorption and the stripping columns. The design procedure proposed has the common feature for the absorber and for the stripper to simultaneously take into account the internal column composition and temperature profiles and the general column performances like the purity or the removal/recovery of the target component.

The columns diameter and packing height together with the absorber feed flowrate were computed for a case where the lean solvent loading value was fixed to 0.3. These results represent the basis for this chapter, which concerns the definition of the whole flowsheet where the interconnection between the absorption and stripping section is considered. Moreover, the flowsheet is completed with the auxiliary equipment and the water-wash section in order to perform the overall economic evaluation of the post-combustion capture plant. This chapter is then divided into two main parts:

1. **Flowsheet Completion**. This part includes the description of the complete flowsheet, the design of the cross heat-exchanger and the water-wash section. For what concerns the cross heat-exchanger, the effect of the rich solvent

C. Madeddu et al., *CO₂ Capture by Reactive Absorption-Stripping*, SpringerBriefs in Energy, https://doi.org/10.1007/978-3-030-04579-1_7

temperature is analyzed to investigate the possibility of a reboiler duty reduction;

2. **Economic Evaluation**. In this part, the capital and the operating costs associated with each piece of equipment are analyzed and computed by means of the Aspen Process Economic Analyzer® (APEA).

## 7.2  Complete Flowsheet

In the following Sections the discussion is made with reference to the updated flowsheet reported in Fig. 7.1. Compared to the flowsheet reported in Chaps. 5 and 6, different pieces of equipment have been added:

- **Water-wash**. It has the function to recover the solvent vaporized in the exhaust stream of the absorber. It consists in an additional packing section above the absorber packing height. It must be noted that in this case the MEA is recovered by physical absorption and that the solvent used is the water from the condenser;
- **Pumps 2-3**. They manage the hot lean solvent exiting from the reboiler (LS-3) and the water recovered from the condenser (W-1);
- **Valves 1-2-3**. They are needed to equalize the pressure of the streams to the operating column to which they are fed;

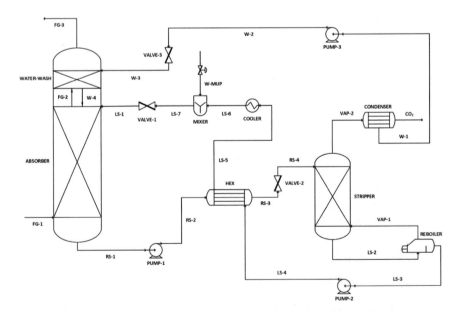

**Fig. 7.1** Industrial $CO_2$ post-combustion capture by reactive absorption-stripping complete plant flowsheet (FG: flue gas; LS: lean solvent; RS: rich solvent; VAP: vapor; W: water)

For what concerns the operating parameters, several differences must be highlighted:

- the outlet pressure of PUMP-1 is risen in order to avoid the partial vaporization of the rich solvent inside the cross heat-exchanger;
- the water recovered from the condenser is not mixed with the lean solvent exiting the cross heat-exchanger. In fact, the water from the condenser is the feed of the water-wash section. The mixing between the two streams happens at the top of the absorption section;
- a water make-up stream (W-MUP) is added to close the water mass balance. This stream is mixed with the lean solvent coming from the cooler;
- the temperature of the rich solvent is risen up to its maximum value. This topic is covered in Sect. 7.3.1.

## 7.3 Cross Heat-Exchanger

After a brief discussion on the role of the stripper feed temperature, this part is dedicated to the procedure for the design of the cross heat-exchanger.

### 7.3.1 Effect of the Stripper Feed Temperature

The stripper feed temperature plays an important role for what concerns the energy consumption. As a matter of fact, when this temperature is increased, the stripping process is favored for two main reasons:

1. in the equilibrium reaction involving the $CO_2$ (Eq. 7.1):

$$CO_{2(aq)} + 2H_2O \rightleftharpoons H_3O^+ + HCO_3^-  \qquad (7.1)$$

the formation of free-$CO_2$ is favored, increasing the amount of free carbon dioxide in the liquid phase. Moreover, the material transfer from the liquid to the gaseous phase starts before the feed enters the stripper. This makes the separation easier and less vapor is needed in the process;

2. the stripping reactions, opposite to the absorption ones, are endothermic. Therefore, a higher feed temperature increases the temperature in the column and, consequently, the rate of the stripping reactions. This leads to a reduction in the energy needed in the process.

| Stripper feed T (K) | Reboiler duty (MW) | Diameter (m) |
|---|---|---|
| Boiling T | 145.7 | 6.93 |
| 373.15 | 106.6 | 6.29 |

**Table 7.1** Variation of the reboiler duty and the column diameter for different values of the stripper feed temperature

From this discussion, the temperature of the rich solvent entering the stripper must be the highest possible. However, the value of this temperature is limited by two constraints:

- the solvent degradation temperature for which, as reported in Chap. 6, a value of 395.15 K was considered according to the work of Alie [1];
- the minimum temperature approach in the cross heat-exchanger, which is fixed to 10 K .

The first constraint is respected by fixing the stripper pressure at the highest value that guarantees a solvent boiling temperature lower than the solvent degradation temperature. This value was defined as 1.8 bar in Chap. 6. Therefore, the second constraint becomes the most tightening one and it can be concluded that the maximum stripper feed temperature allowed is the highest one that ensures the respect of the minimum temperature approach in the cross heat-exchanger. In the analyzed case, this temperature is found to be 373.53 K. For what concerns the variation of the reboiler duty and the column diameter with the stripper feed temperature, Table 7.1 compares the results obtained when this temperature is at its boiling point and at its maximum value.

From the analysis of Table 7.1, it is evident the positive effect derived from the increase of the stripper feed temperature at the maximum allowed value. In fact, moving from the boiling point to the maximum value, a reduction of about 27% in the value of the reboiler duty is observed. For what concerns the column diameter, a reduction of about 0.6 m is obtained. The maximum stripper feed temperature is the one considered for the rest of the chapter.

## 7.3.2  Cross Heat-Exchanger Design

The cross heat-exchanger represents the connection between the absorption and the stripping sections. In fact, here the rich solvent from the bottom of the absorber is heated using the sensible heat of the lean solvent coming from the reboiler. The heat-exchanger should be designed in order to maximize the exchange between the two streams. As demonstrated in the previous Section, the rich solvent must be sent to its maximum value to obtain the minimum energy consumption in the reboiler. However, at the stripper pressure, i.e., 1.8 bar, the rich solvent, due to the presence of $CO_2$, starts to flash before reaching the maximum temperature. In this way, a

**Table 7.2** Design results for the cross heat-exchanger

| Parameter | Value |
|---|---|
| Heat Duty (MW) | 112.6 |
| Average heat transfer coefficient [W/(m$^2$ K)] | 338.34 |
| Exchange area (m$^2$) | 28,892 |

two-phase flow would be present inside the heat-exchanger and acid breakout and corrosive phenomena happen [2, 3]. Therefore, to avoid these problems, the outlet pressure from PUMP-1 is increased to 7 bar, which ensures the absence of $CO_2$-flashing inside the cross heat-exchanger. At this point, the temperature of the rich solvent exiting the heat-exchanger was set at 380.15 K. In this way, the stripper feed temperature is at the maximum value of 373.53 K defined in Sect. 7.3.1 after VALVE-2 in Fig. 7.1.

The design of the heat-exchanger was done to evaluate the exchange area, which is the most important parameter for the evaluation of the heat-exchanger cost. According to the well-known design equation for a heat-exchanger (Eq. 7.2):

$$Q = UA\Delta T_{lm} \tag{7.2}$$

the duty, $Q$, the overall heat transfer coefficient, U, and the logarithmic mean temperature difference, $\Delta T_{lm}$, are needed to evaluate the heat transfer area, $A$.

In particular, the Aspen Exchanger Design and Rate® (Aspen EDR) was used to evaluate the average heat transfer coefficient. Then, the *Shortcut* mode within Aspen Plus® was used to evaluate the exchange area. Table 7.2 resumes the results for the design of the cross heat-exchanger.

## 7.4   Water-Wash Section

The large amount of heat generated in the absorption process due to the exothermic reactions leads to the vaporization of part of the solvent in the exhaust gas. This loss has consequences at an environmental and an economic level. In fact, the amine cannot be emitted in the atmosphere due to its harmful potential for both the environment and the human beings. Regulations exists on the maximum amount of solvent that can be released in the environment and this information is usually expressed as a threshold limit value. In particular, for MEA the TLV-TWA (Threshold Limit Value — Time-Weighted Average) is 3 ppm, according to the Material Safety Data Sheet from the Dow Chemical Company [4]. Moreover, the solvent cost is typically high and its loss might have an important influence in the economics of the plant. For this reason, a solvent recovery section is needed to avoid the loss of MEA with the exhaust gas.

As already mentioned in Sect. 7.2, the water-wash consists in a physical absorption operation. This is done adding a packing section above the absorber's one and using the water from the condenser as solvent. The design procedure for a physical absorption process recalls the one reported in Chap. 5, i.e., computation of the minimum solvent flow rate with an infinite packing height followed by the evaluation of the effective solvent flow rate and column dimensions (packing height and diameter). In this specific case, part of these unknowns are already determined. In fact, the effective solvent flow rate is represented by the water stream recovered from the condenser, while the diameter corresponds to the absorber diameter, since the water-wash section is inside the same vessel. At this point, the only element to be computed is the water-wash packing height, which was done by means of a sensitivity analysis. In particular, the target was to find the packing height that ensures to have a concentration of MEA in the exhaust gas less than 1 ppm. This choice is made to safely respect the limit imposed by the TLV-TWA for MEA. This target also ensures an almost complete recovery of the solvent. It must be noted that in Aspen Plus®, to simplify the calculations, the absorber and the water-wash section were treated as two separated columns with the same diameter and packing type. Furthermore, the mixing between the water from the bottom of water-wash section and the lean solvent from the cooler was modeled using the MIXER model. This modeling approach is showed in the Aspen Plus® flowsheet reported in Fig. 7.2.

Table 7.3 resumes the results for the water-wash section design.

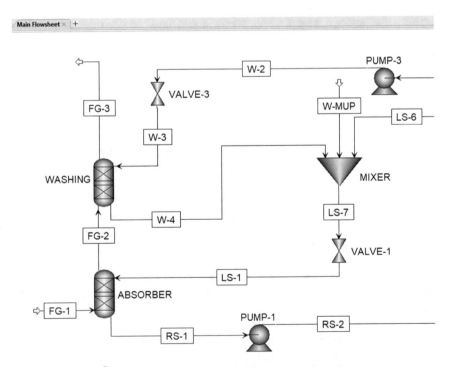

**Fig. 7.2** Aspen Plus® flowsheet for the absorber and the water-wash section

**Table 7.3** Design results for
the water-wash section

| Parameter | Value |
|---|---|
| Inlet MEA flow rate (kg/h) | 258.7 |
| Outlet MEA concentration (ppm) | 0.918 |
| MEA recovery (%) | 99.83 |
| Packing height (m) | 3.5 |

**Fig. 7.3** Water-wash section
MEA composition profile in
the gaseous phase

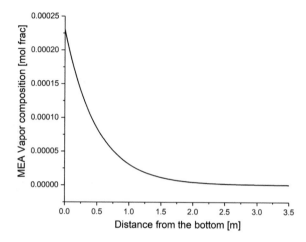

The correct operation of the designed water-wash section is demonstrated by the
MEA vapor molar fraction profile, reported in Fig. 7.3.

As it is possible to notice from the analysis of Fig. 7.3, the MEA concentration
in the gas is reduced along the packing height up to almost zero when the top of the
water-wash section is reached.

## 7.5   Auxiliary Equipment

The last elements in the flowsheet are represented by the cooler, the pumps and the
valves. In particular, the cooler is needed to reduce the temperature of the lean
solvent before it enters the top of the absorber. On the other hand, the pumps are
needed to move the streams across the plant, while the valves are used to reduce the
feed streams pressure to the value required in their respective columns. The design
of these items is not reported since this is outside the scope of this chapter.
However, the costs associated to the auxiliary equipment are taken into account
using the Aspen Process Economic Analyzer® (APEA).

**Table 7.4** Overview of all the streams involved in the complete process flowsheet in Fig. 7.1

| Stream | Flow rate (tonne/h) | Temperature (K) | Pressure (bar) | Vapor fraction (%) | $CO_2$ (mol frac) | $H_2O$ (mol frac) | MEA (mol frac) | $N_2$ (mol frac) |
|---|---|---|---|---|---|---|---|---|
| *Gas/Vapor* | | | | | | | | |
| FG-1 | 540 | 313.15 | 1 | 1 | 0.1412 | 0.069 | 0 | 0.7898 |
| FG-2 | 482.29 | 334.45 | 1 | 1 | 0.014 | 0.2017 | 0.0002 | 0.7841 |
| FG-3 | 480.722 | 332.96 | 1 | 1 | 0.0141 | 0.1985 | $3.94 \times 10^{-7}$ | 0.7874 |
| VAP-1 | 166.94 | 391.72 | 1.8 | 1 | 0.05 | 0.9476 | 0.0024 | 0 |
| VAP-2 | 144.97 | 372.24 | 1.8 | 1 | 0.495 | 0.5046 | 0.0003 | 0.0001 |
| $CO_2$ | 103.45 | 308.15 | 1.8 | 1 | 0.9679 | 0.032 | 0 | 0.0001 |
| *Liquid* | | | | | | | | |
| LS-1 | 2071.44 | 308.15 | 1 | 0 | 0.0363 | 0.8426 | 0.1211 | 0 |
| RS-1 | 2172.24 | 320.08 | 1 | 0 | 0.0618 | 0.8201 | 0.1181 | $4.54 \times 10^{-6}$ |
| RS-2 | 2172.24 | 319.93 | 7 | 0 | 0.0618 | 0.8201 | 0.1181 | $4.54 \times 10^{-6}$ |
| RS-3 | 2172.24 | 380.15 | 7 | 0 | 0.0618 | 0.8201 | 0.1181 | $4.54 \times 10^{-6}$ |
| RS-4 | 2172.24 | 373.53 | 1.8 | 0.01 | 0.0579 | 0.8229 | 0.1192 | $4.54 \times 10^{-6}$ |
| LS-2 | 2194.21 | 389.7 | 1.8 | 0 | 0.0386 | 0.8482 | 0.1132 | 0 |
| LS-3 | 2027.26 | 391.72 | 1.8 | 0 | 0.0374 | 0.8379 | 0.1247 | 0 |
| LS-4 | 2027.26 | 391.81 | 6 | 0 | 0.0374 | 0.8379 | 0.1247 | 0 |
| LS-5 | 2027.26 | 331.29 | 6 | 0 | 0.0374 | 0.8379 | 0.1247 | 0 |
| LS-6 | 2027.26 | 308.15 | 6 | 0 | 0.0374 | 0.8379 | 0.1247 | 0 |
| W-MUP | 44.17 | 298.15 | 6 | 0 | 0 | 1 | 0 | 0 |
| LS-7 | 2071.44 | 308.15 | 6 | 0 | 0.0363 | 0.8426 | 0.1211 | 0 |
| W-1 | 41.54 | 308.15 | 1.8 | 0 | 0.0015 | 0.0006 | 0.9979 | 0 |
| W-2 | 41.54 | 308.24 | 6 | 0 | 0.0015 | 0.0006 | 0.9979 | 0 |
| W-3 | 41.54 | 308.24 | 1 | 0 | 0.0011 | 0.0006 | 0.9983 | 0 |

## 7.6  Process Streams Characterization Overview

At this point, before the evaluation of the plant costs, an overview of the streams involved in the complete flowsheet showed in Fig. 7.1 is presented in Table 7.4. In particular, for each stream the flow rate, temperature and pressure are reported together with the composition of the main components.

## 7.7  Economic Analysis and Evaluation

The objective of this second part of the chapter is the evaluation of the total annual cost (TAC) of the plant. The TAC is given by the sum of the annualized capital and the operating costs [5, 6], as reported in Eq. 7.3:

$$TAC = \frac{CC}{Payback\ Period} + OC \tag{7.3}$$

In particular, the first term is referred to the purchase costs for the equipment (Purchased Costs) and the equipment installment (Installed Costs). On the other hand, the operating costs corresponds to the costs for the utilities, i.e., steam, cooling water, electricity. For the economic calculations, Aspen Plus® can be linked to the Aspen Process Economic Analyzer®, which is the software from AspenTech dedicated to the costs estimation. Some pieces of information must be highlighted:

- the capital costs of each piece of equipment vary depending on the material used for their construction. The materials were chosen according to the report of the *Asian Development Bank (ADB)* published on the official site of the Global CCS Institute [7];
- two outputs are given by the APEA for the capital costs: the purchased costs and the installed costs. The first item represents the costs for the purchase of the bare equipment, while the latter represents the sum of the bare equipment costs and the costs for the equipment installation. In this book, the installed costs are reported as representative of the capital costs;
- the capital costs are linearly depreciated over an assumed payback period of 10 years;

In the next sections, for every piece of equipment the dimensions involved for the evaluation of the annualized capital costs and the results are reported in a tabular form. Then, the costs for the utilities, i.e., the operating costs, are computed. In the end, the total annual costs are presented and commented.

### 7.7.1 Packed Columns

In this Section, the absorber + water-wash section and the stripper are considered. In APEA, together with the column dimensions, it was necessary to define:

- Application, i.e., absorption/stripping;
- Shell material;
- Packing material;
- Vessel tangent to tangent height, which is the column total height, given by the sum of the packing height, redistributors, bottom and top disengagements;
- Number of packed sections, which was chosen assuming a maximum packed section height of 5 m, in agreement with the plant reported in the work of Razi et al. [8];

The inputs and the results are reported in Tables 7.5 and 7.6.

**Table 7.5** Dimensions and annualized capital costs for the absorption + water-wash column

| Variable | Value |
|---|---|
| Application | ABSORB |
| Absorption packing height (m) | 29.8 |
| Water-Wash packing height (m) | 3.5 |
| Total packing height (m) | 33.3 |
| Total column height (m) | 37.6 |
| Number of packed section | 7 |
| Diameter (m) | 9.67 |
| Packing volume (m$^3$) | 2445.6 |
| Shell material | Carbon Steel—A516 |
| Packing material | 304 Stainless Steel—M76YB |
| Annualized capital costs (k\$/y) | 1323.62 |

**Table 7.6** Dimensions and annualized capital costs for the stripping column

| Variable | Value |
|---|---|
| Application | STR-RB |
| Total packing height (m) | 7 |
| Total column height (m) | 11.3 |
| Number of packed section | 2 |
| Diameter (m) | 9.67 |
| Packing volume (m$^3$) | 2445.6 |
| Shell material | A516 |
| Packing material | M76YB |
| Annualized capital costs (k\$/y) | 157.07 |

## 7.7.2  Heat-Exchangers

All the pieces of equipment dealing with heat exchange, i.e., cross heat-exchanger, reboiler, condenser and cooler, are the object of this Section. The necessary input in APEA for heat exchange equipment are:

- Total exchange area;
- Shell material;
- Tube material;
- Number of exchangers, which is calculated by the Aspen EDR;
- TEMA type [9].

It must be highlighted that the reboiler, condenser and cooler exchange area are evaluated by the APEA software using the Interactive Sizing tool. The inputs and the results are reported in Table 7.7.

**Table 7.7** Dimensions and annualized capital costs for the heat-exchangers

| Variable | Cross HEX | Reboiler | Condenser | Cooler |
|---|---|---|---|---|
| Total exchanger area (m²) | 28892.6 | 3711.1 | 1341.65 | 4123.91 |
| Number of exchangers | 30 | 4 | 2 | 5 |
| Shell material | A285C | A285C | A285C | A285C |
| Tube material | A214 | A214 | A214 | A214 |
| TEMA type | BEM | BKU | BEM | BEM |
| Annualized capital costs (k$/y) | 1003.19 | 114.85 | 40.78 | 107.78 |

## 7.7.3 Pumps

The last category of equipment to be treated is represented by the pumps. Though, as reported in Sect. 7.5, the design of these equipment is not treated in this book, the APEA software is still able to perform an estimation of the capital costs. In particular, the inputs to be defined are:

- Liquid flow rate managed;
- Outlet pressure;
- Casing material;
- Type of pump.

The inputs and the results are reported in Table 7.8.

## 7.7.4 Utilities

Utilities are needed in the plant to subtract heat (cooling water), supply heat (steam), supply power (electricity). Moreover, the treated make-up water is considered in this Section since it represents an operating cost. The specific costs for the utilities are taken from the literature and reported in Table 7.9.

### 7.7.4.1 Cooling and Make-up Water

For what concerns the make-up, as reported in Table 7.4, the need for 44.17 tonne/h of filtered and softened water, correspond to a cost of 119.95 k$/y. On the other hand, Table 7.10 resumes the results for the cooling water.

**Table 7.8** Dimensions and annualized capital costs for the pumps

| Variable | PUMP-1 | PUMP-2 | PUMP-3 |
|---|---|---|---|
| Liquid flow rate (m³/s) | 0.53 | 0.55 | 0.0116 |
| Outlet pressure (bar) | 7 | 6 | 6 |
| Casing material | CS | CS | CS |
| Pump type | CENTRIF | CENTRIF | CENTRIF |
| Annualized capital costs (k$/y) | 34.96 | 33.51 | 4.12 |

**Table 7.9** Specific costs for the utilities

| Parameter | Value | Reference |
|---|---|---|
| Cooling water ($/tonne) | 0.082 | [10] |
| Filtered and softened water ($/tonne) | 0.305 | [6] |
| LP-Steam at 144 °C ($/tonne) | 13 | [10] |
| Electricity ($/kWh) | 0.1 | [10] |

**Table 7.10** Utilization and costs for the cooling water

| Variable | Condenser | Cooler | Total |
|---|---|---|---|
| Duty (MW) | 30.81 | 42.29 | 73.1 |
| Utilization (tonne/h) | 1772.12 | 1819.58 | 3591.7 |
| Total cost (k$/y) | 1273.83 | 1307.93 | 2581.76 |

**Table 7.11** Utilization and costs for the LP-Steam

| Variable | Reboiler |
|---|---|
| Duty (MW) | 106.58 |
| Utilization (tonne/h) | 179.84 |
| Total cost (k$/y) | 20493.9 |

**Table 7.12** Utilization and costs for the electricity

| Variable | PUMP-1 | PUMP-2 | PUMP-3 | Total |
|---|---|---|---|---|
| Utilization (kW) | 372.45 | 270.65 | 7.986 | 651.08 |
| Total cost (k$/y) | 326.49 | 237.25 | 7 | 570.74 |

#### 7.7.4.2   Low-Pressure Steam

The costs related to the consumption of the LP-steam in the reboiler are reported in Table 7.11.

#### 7.7.4.3   Electricity

For what concerns the electricity used for the pumps, the results are showed in Table 7.12.

### 7.7.5   Total Costs Evaluation

Once the annualized capital costs for each piece of equipment and the operating costs related to each utility are evaluated, it is possible to compute the total annual costs of the plant using Eq. 7.3. The results are reported in Table 7.13.

**Table 7.13** Total costs of the plant

| Variable | Reboiler |
|---|---|
| Annualized capital costs (k$/y) | 2819.88 |
| Operating costs (k$/y) | 23646.36 |
| Total annual costs (k$/y) | 26466.24 |

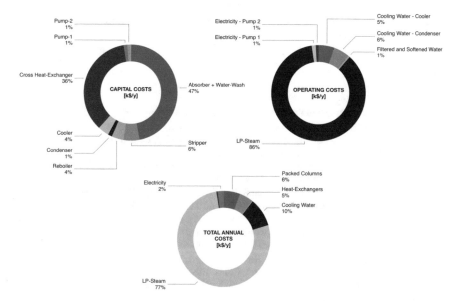

**Fig. 7.4** Overview of the percentage distribution among annualized capital costs and operating costs

An overview of how the different pieces of equipment and the utilities are distributed among the costs is reported in Fig. 7.4, where the annualized capital, the operating and the total annual costs are presented in a donut-chart.

Different observations can be made from the analysis of Fig. 7.4:

- The absorption + water-wash column alone is responsible for the 47% of the total annualized capital costs, since it is the largest equipment in the whole capture plant, followed by the cross heat-exchangers train (36%). The annualized capital costs are almost equally divided between the packed columns (50%) and the heat-exchange equipment (45%);
- the operating costs are dominated by the consumption of LP-steam in the reboiler, which represents almost the 90% of the variable expenses. The utilization of cooling water constitutes largely the remaining of the variable costs;
- the trend noticed in the operating costs analysis is repeated for what concerns the total annual costs. In fact, the majority of the overall plant costs is related to the consumption of LP-steam, which is responsible for almost the 80% of the expenses. It is notable that the total annual capital costs represents only the 11% of the total annual costs, roughly more than the costs for the consumption of cooling water.

From what was said above, it emerges that the operating costs reduction, with particular focus on the reboiler duty, is the key for the optimization of a $CO_2$ reactive absorption-stripping plant. For example, the use of solvents that require less heat for their regeneration could be a possibility [11]. At the same time, process intensification and process integration techniques can be applied [12, 13]. All of these efforts are going to play a major role in promoting this technology a systematic solution for the reduction of the industrial $CO_2$ emissions.

# References

1. Alie CF (2004) $CO_2$ capture with MEA: integrating the absorption process and steam cycle of an existing coal-fired power plant. Dissertation, University of Waterloo
2. DuPart MS, Bacon TR, Edwards DJ (1993) Understanding corrosion in alkanolamine gas treating plants Part 1. Hydrocarbon Process 72:75–80
3. DuPart MS, Bacon TR, Edwards DJ (1993) Understanding corrosion in alkanolamine gas treating plants Part 1. Hydrocarbon Process 72:89–94
4. The Dow Chemical Company Material Safety Data Sheet. Monoethanolamine. MSDS#:1592 (Online). The Dow Chemical Company, Midland MI. 17/06/2003
5. Sinnott RK (2005) Coulson & Richardson's chemical engineering, vol 6—Chemical engineering design. Elsevier Butterworth-Heinemann
6. Peters MS, Timmerhaus KD, West RE (2003) Plant design and economics for chemical engineers. McGraw-Hill Education, New York
7. Beijing Jiaotong University and North China Electric Power University (2014) People's Republic of China: study on carbon capture and storage in natural gas-based power plants. ADB Technical Consultant's Report
8. Razi N, Svendsen HF, Bolland O (2013) Validation of mass transfer correlations for $CO_2$ absorption with MEA using pilot data. Int J Greenhouse Gas Control 19:478–491
9. Kakaç S, Liu H, Pramuanjaroenckij A (2012) Heat exchangers—selection, rating, and thermal design. CRC Press, Boca Raton
10. Rév E, Emtir M, Szitkai Z et al (2001) Energy savings of integrated and couple distillation systems. Comput Chem Eng 25(1):119–140
11. Tobiesen FA, Haugen G, Hartono A (2018) A systematic procedure for process energy evaluation for post combustion $CO_2$ capture: case study of two novel strong bi-carbonate forming solvents. Appl Energy 211:161–173
12. de Miguel Mercader F, Magneschi G, Fernander ES et al (2012) Integration between a demo size post-combustion $CO_2$ capture and full size plant. An integral approach on energy penalty for different process options. Int J Greenhouse Gas Control 11S:S102–S113
13. Jung J, Jeong YS, Lee U et al (2015) New configuration of the $CO_2$ capture process using aqueous monoethanolamine for coal-fired power plants. Ind Eng Chem Res 54(15):3865–3878

# Chapter 8
# Conclusions

The modeling of reactive absorption-stripping systems is known to be a challenging task, and even if sophisticated process simulators are available, a high-level knowledge of the fundamental transfer and kinetics phenomena is essential.

In the first part of the book it was proved that the basic settings of the $RadFrac^{TM}$ model—Rate-Based mode available in Aspen Plus® is not enough to develop a good model. In particular, considering the $CO_2$ capture by means of reactive absorption-stripping process using MEA as solvent, a new approach to the process modeling based on the analysis of the system fluid dynamics was introduced.

This analysis consists in the evaluation of the dimensionless Peclet number to examine the possible effect of the axial dispersion followed by the definition of a proper number of segments to obtain the correct solution of the resulting system of algebraic equations from a numerical point of view. Once the correct solution is found, it is possible to investigate the possible effect of the backmixing generated by the countercurrent.

The model developed was firstly applied to the absorption section of two pilot-plant facilities with different packing and operating conditions. Following the analysis of the system fluid dynamics and the obtainment of a correct numerical solution of the system of equations, the model was able to describe correctly each set of experimental data, particularly for what concerns the typical absorber temperature bulge, independently on its position and magnitude. Then, the stripping section of two other pilot-plant facilities was considered. In this case, two sets of degrees of freedom were defined for the columns in order to test the model with respect to the evaluation of the internal profiles and the reboiler duty. With respect to the reboiler duty, it was found that using a correct model of the process it was possible to reduce the error between the experimental value and the model result from 21.4% (10 segments) to 0.6% (90 segments). Even for the stripper, the model was validated for every set of experimental data considered. The model developed is a useful tool for the extension of the system at a dynamic level and for the subsequent implementation of a control system.

© The Author(s), under exclusive license to Springer Nature Switzerland AG 2019
C. Madeddu et al., *CO₂ Capture by Reactive Absorption-Stripping*,
SpringerBriefs in Energy, https://doi.org/10.1007/978-3-030-04579-1_8

After the validation, the model was used to analyze the design procedure of an industrial scale plant. Initially the design of the absorber was considered. A two-step procedure consisting in the evaluation of the minimum number of absorption units and minimum solvent flow rate with an infinite packing height column followed by the evaluation of the effective solvent flow rate and column dimensions was implemented. Furthermore, the operating conditions for which the absorber does not show any isothermal zone were defined on the basis of the liquid temperature profiles analysis. The results from the absorber design were used to analyze the design of the stripper. Since the reboiler duty is needed only to reverse the absorption reactions only, an alternative plant configuration without reflux was adopted to avoid an unnecessary energy consumption to heat the water from the condenser. Furthermore, an alternative approach was proposed for the evaluation of the stripper packing height referring to the maximum packing height at which the temperature gradient is always higher than 1 K/m. In the end, the interconnection between the two main sections was introduced by means of the cross heat-exchanger design. In this context, it was also showed that the stripper feed temperature must be sent at the maximum allowed value, which is constrained by the respect of the minimum temperature approach in the cross-heat exchanger. Furthermore, the flowsheet was completed adding the auxiliary equipment and designing the water-wash section for the solvent recovery. Then, the total annual costs associated with the plant were evaluated as the sum of the total annualized capital costs and the operating costs using the Aspen Process Economic Analyzer®. In particular, it was identified the reduction of the operating costs as the crucial point for the optimization of the process, in order to make it a systematic solution in the field of the industrial $CO_2$ capture.

The design and the economic evaluation procedures reported represent a valid basis for a complete process design of an industrial system.

Printed in the United States
By Bookmasters